2023 젊은 건축가상
의미, 무용, 태도

2023 Korean Young Architect Award
Meaning, Futility, Attitude

김진휴 남호진	김남건축
김영수	모어레스 건축
서자민	아지트 스튜디오

Jinhyu Kim Hojin Nam	KimNam Architects
Youngsoo Kim	More Less Architects
Jamin Seo	AGIT STUDIO

일러두기
수상 건축가 게재 순서는 건축사사무소 이름을 기준으로 가나다 순을 따랐다. 모든 원고는 국문과 영문을 병기해 수록했다.

Editor's Note
The winners' section were arranged in Korean alphabetical order of their practice names.
All texts were written both in Korean and English.

4—5		편집자의 글	Editor's Letter
6—67		김남건축	KimNam Architects
		김진휴, 남호진	Jinhyu Kim, Hojin Nam
	68—83	짚어보기	Review
		남성택	Sungteag Nam
84—145		모어레스 건축	More Less Architects
		김영수	Youngsoo Kim
	146—155	짚어보기	Review
		박진호	Jinho Park
156—219		아지트 스튜디오	AGIT STUDIO
		서자민	Jamin Seo
	220—227	짚어보기	Review
		양수인	Sooin Yang
228—233		심사총평	General Review
		이민아	Minah Lee

편집자의 글

매년 한국의 젊은 건축가를 대상으로 시행하는 이 상은 도록 발간을 통해 수상자들이 자신의 '건축적 세계를 피력'할 수 있는 기회를 제공하고 있다. 이를 계기로 수상자들은 지나온 길을 되돌아보고 내 건축의 시작은 어디인지, 무엇에 천착하는지, 지금은 어디에 머물러 있는지, 궁극에 닿고자 하는 지향점은 무엇인지를 깊이 생각하게 된다. 이 같은 이유로 이 책은 단순히 젊은 건축가를 홍보하는 것 이상의 의미를 갖는다.

모든 수상자가 기획 단계부터 참여하여 함께 머리를 맞대고 책에 무엇을 담을지, 구성은 어떻게 할지, 리뷰를 수록한다면 필자는 누구로 할지를 논의하는 것도 이제는 이 책의 전통이 됐다. 외부 편집자가 참여하여 실험적인 기획서를 제안하기도 하지만, 이 또한 수상자들의 동의를 전제로 한다. 기실 편집자에게 요구되는 건 조율의 역할일 뿐, 중요한 결정은 수상자들의 몫인 셈이다.

건축가의 작업을 소개하는 대다수의 건축 책이 그렇듯, 지금까지 젊은 건축가상의 도록은 대체로 (수상의 소회를 겸하는) 건축가의 글과 주요 프로젝트를 사진과 도면 중심으로 소개하고 그들의 작업을 잘 전달해줄 수 있는 외부 필자의 리뷰를 수록하는 형식이었다. 조금 다른 구조를 찾아본다면 대담 혹은 집담을 담아낸 정도이거나, 이보다 파격적이라면 각자의 건축적 사유를 몇 개의 주제어로 담담히 풀어낸 사례가 있다.

올해의 수상자들이 선택한 수상집의 모습은 이중 마지막 형식이다. 바쁜 시간을 쪼개 원고지 7~80매에 달하는 글을 쓰는 일이 결코 녹록하지는 않겠지만, 건축가로서의 문제의식과 관심사 등을 이야기로 풀어내는 것은 "건축설계와 글쓰기는 주제, 구조, 공간, 디테일을 다 갖추고 있는 점에서 유사성이 있고, 특히 건축가를 선정하는 상이라 사유가 언어로 옮겨지는 단계에 개인적으로 큰 관심을 두었다"는 심사평의 한 대목에 부합하는 형식이기도 하다.

이에 따라 이 책은 대여섯 개의 주제어 아래 일상, 관심사, 지향점, 프로젝트, 문제해결 방식 등등 다양한 내용을 담은 수상자들의 에세이로 묶였다. 이 에세이들은 최소한의 교정만으로 글쓴이의 문체를 가급적 있는 그대로 유지하여 각자의 개성이 드러나도록 했다. 또 각 팀의 글 바로 뒤에는 수상자를 따뜻한 시선으로 응원하는 선배 건축인의 리뷰를 붙였다.

확실히 이러한 책은 사진과 도면 중심의 프로젝트 소개집에 비해 어려운 형식임은 분명하다. 그러나 글을 쓰고 퇴고를 거치는 동안 건축가의 내면은 더욱 단단해지고, 독자들은 글을 읽으며 충분히 올해의 수상자를 알아갈 수 있다는 점에서 권할 만한 방식이라 생각한다.

The Editor's Letter

The Korean Young Architect Award, held annually, provides a valuable opportunity for winners to express their opinions on their own architectural philosophies through a book. In other words, this event serves as a chance for them to look back on the paths they have walked and contemplate the origins of their architecture, the subjects they have explored, their current positions, and their ultimate goals. In this vein, this book is more than just a platform for promoting young architects.

Traditionally, all the winners participate in planning the book, brainstorming the content it should contain, how it should be structured, and who should write reviews. An outside editor often plays a role in this process, proposing an experimental plan, but these proposals are subject to approval by the winners. In fact, the editor's role is primarily that of a coordinator, with the winners making important decisions.

Like most art books introducing the process of architects creating their work, this book also comprises architects' thoughts (their feelings about winning the award), key project photos & blueprints, and an outsider's review. This often also includes conversations or group discussions, or keyword-based special discourses on architectural ideas.

The book for this year is organized in the latter form out of those mentioned above. It will not be easy to take time out of their busy schedules to write 70 to 80 pages on 200 square writing paper. However, discussing their interests and critical content as architects in written form is deemed to correspond to the following comment: "Architectural design and writing share similarities, as both involve themes, structures, spaces, and details. Personally, I was particularly intrigued by how applicants translated their thoughts into language because this award is presented to architects themselves."

Against this backdrop, this book consists of winners' essays containing various information such as their daily lives, areas of interest, orientations, projects, and problem-solving methods under five to six keywords. These essays have been revised or edited as minimally as possible to fully showcase the authors' individuality. Essays by teams are also followed by reviews authored by senior architects who warmly support the young winners.

It is true that compiling these books is more challenging than presenting blueprint-based project introductions. However, such books are recommended because the process of writing and revising the essays strengthens the architects mentally, and readers gain a better understanding of the winners as they peruse the essays.

KimNam Architects

김남건축 | Jinhyu Kim, Hojin Nam

김진휴 남호진

건축사사무소 김남은 2014년 스위스의 산골 마을에서 시작된 건축설계사무소이다. 2015년부터 서울에서 활동하며 작업을 이어가고 있다. 건축에 존재하는 다양한 가치와 관점의 존재를 중시하며, "어제 옳은 것이 오늘 틀릴 수 있다"는 시각으로 의심하고 다시 그린다.

김진휴는 서울대학교 건축학과를 졸업하고 미국 예일대학교에서 석사학위를 받았다. 스위스의 Herzog & de Meuron, 일본의 SANAA, 미국의 SO-IL에서 건축실무를 익혔다. 서울대학교, 한양대학교 등에서 강의를 했다.

남호진은 이화여자대학교 건축학과를 졸업하고 미국 예일대학교에서 석사학위를 받았다. 미국의 Pelli Clarke Pelli Architects, 스위스의 Herzog & de Meuron, (주)남산 에이엔씨 종합건축사사무소에서 실무를 익혔다. 2023년부터 이화여자대학교 겸임 교수로 재직 중이다.

KimNam Architects started in a mountain village in Switzerland in 2014 and launched in Seoul in 2015. The architectural firm emphasizes a great diversity of values and perspectives that exist in architecture, and does not hesitate to redraw designs, if necessary, from the perspective that "what was right yesterday can be wrong today."

Jinhyu Kim graduated from Seoul National University with a bachelor's degree in architecture and received a master's degree from Yale University in the United States. He built his architectural career with Herzog & de Meuron in Switzerland, SANAA in Japan, and SO-IL in the United States. He has taught at Seoul National University, Hanyang University, and other institutions.

Hojin Nam graduated from Ewha Womans University with a bachelor's degree in architecture and received a master's degree from Yale University in the United States. She built her architectural career with Pelli Clarke Pelli Architects in the U.S., Herzog & de Meuron in Switzerland, and Namsan A&C in Korea. She has been serving as an adjunct professor at Ewha Womans University since 2023.

The Taboo of Balance

균형이라는 금기

(김남건축은) 모두가 회피하는 건축의 아름다움을 읊었다. (중략) 즉 눈에 보이지 않는 것으로부터 정의하고 치열하게 구조화하여 시공자의 수고, 사용자의 기쁨, 건축가 스스로의 검열이 동반될 때 비로소 아름다움에 이르는 길에 당도했음을 성찰한다. 그리고 그러한 태도로 "결국 건축가는 무엇에 헌신하는 사람인가"라는 질문을 강력히 던지고 있다.

두 사람의 건축은 인간과 주변을 사색하는 일에서 시작하여 '아름다움'보다 실은 훨씬 더 어려운 '윤리적' 건축의 실천을 위해 부단히 노력 중임을 목격했다.

— 심사평 중

(KimNam Architects) sang the beauty of architecture that everyone avoids. (Omitted)In other words, they reflected on how they were able to reach the path to beauty only when making definitions and rigorous structuring with the invisible, accompanied by the constructor's efforts, the user's joy, and the architect's self-censorship. In this attitude, they are strongly posing the question, "Ultimately, what is it that an architect dedicates themselves to?"

We witnessed the duo's architecture begins with contemplating human beings and the surroundings and is tirelessly striving for practicing "ethical" architecture, which is actually far more challenging than realizing "beauty."

— From Jury's comment

두 편의 에세이　50

남 이야기　38

그러기로 하는 이유　28

시간의 의미　18

오늘 우리는　10

Today's Portrait	11
The Meaning of The Time	19
The Reason for Doing So	29
This is Not My Story	39
Two Essays	51

오늘 우리는

"나 내일부터 출근 안 할래. 내 이름 빼고 '김건축' 해."
또 시작이다. 가만 보면 한 4~5개월마다 한 번씩 저러는 것 같다. 익숙해질만도 한데 나는 또 가슴이 철렁한다. 본인이 제안한 것에 반대했다고 정색을 하고 화를 내는 것인데, 나는 황당하고 억울하다. 내가 만들어 둔 것에 대해서 본인은 "이럴 거면 뭐하러 그리지?" "여기는 구려도 되나 보지?" 하는 식의, 차라리 쌍욕을 듣는 게 낫겠다 싶은 발언을 서슴지 않고 해댄다. 그러면서 내가 단점을 지적했다고 저런 반응이다. 매번은 아니고 몇달에 한 번씩.

어찌됐건 벌어진 일, 나는 또 그녀를 설득한다. 내가 단점을 언급할 수밖에 없었던 맥락과 그렇게 크게 나쁘다는 의미가 아니었다는 사실을 설명한 뒤, '김남'건축에 왜 '남'의 의견이 꼭 필요한지 역설한다. 내가 도면을 더 그리라고 하냐, 마감 시간을 맞추라고 하냐, 자꾸 나한테 이러지 말어……. 결혼한 지 10년도 더 된 아내에게 주기적으로 통사정하는 내 자신의 초라함에 서글퍼하다가 '나는 아마도 전생에 나라 팔아먹은 나쁜 놈이었을 것이야' 하는 생각을 하면서 잠이 든다.

미완성 도면의 목록을 작성하고, 현장과 발주처에서 온 전화를 받고, 경리 업무까지 처리하다 보면 진이 쭉쭉 빠진다. 건축적 이상은 개뿔, 마음 편히 잠이나 자고 싶다. 그 상태에서 눈이 빠져라 디테일을 그리다가 '이거 시공사가 엄청 싫어할텐데. 그냥 좀 쉬운 걸로 타협할까' 싶을 때 남소장을 부른다. 남소장, 이 부분 좀 봐봐. 여기가 말이야, 어쩌구 저쩌구…….

"여기는 구려도 되나 보지? 아무한테도 안 보이는 그런 곳이야?"
"……"

왼쪽.
Mimosa

오른쪽.
이의동 상가주택
Curve Cut Sharply

The Taboo of Balance

Today's Portrait

"I quit. Just go ahead and change the company's name to Kim Architects."

There she goes again. It's a cycle that repeats every four to five months. It's about time I get used to it, but my heart sinks again. With an annoyed look on her face, she feels completely offended simply because I opposed to her suggestion. But I think she's being unfair and ridiculous. When it comes to my ideas, she's not hesitant to say like this: "If you're going to do this, why even bother drawing?" or "So this area can be crappy?" Why not just go ahead and curse me. All because I made a comment. Not all the time though. Just every few months.

Anyhow, here I am persuading her again. I first explain why I "had" to point that out but of course, never meant it in a bad way, and I finish my sentence off with the reason why "Nam" is absolutely necessary at KimNam Architects. You know I don't ask you for more floor plans or push you to meet deadlines, please don't do this to me....... Pitying myself for having to regularly beg my wife of more than 10 years, I fall asleep blaming my past life karma.

Making a list of unfinished drawings, answering calls from job sites and clients, and even processing accounting tasks can be quite exhausting. Heck with architectural ideals. I just want to sleep. But in reality, I'm filling in the details, and a part of me wonders if I should just compromise on something a little easier, thinking, "The contractor is going to hate this so much." This is when I call upon Nam. Take a look at this part. This here, blah blah.......

"So this area can be crappy? Is this like an abandoned spot?"

"........"

옳다고 믿고 있던 사실이 틀렸다는 것을 알게 되는 것은 발전하기 위해서 매우 중요한 일이다. 내가 어떤 의견을 말했을 때 그에 대한 반론을 듣고 내가 수용하게 될 때 나는 그 커뮤니케이션에 대해 보람을 느낀다. (남소장은 꼭 그런 것 같지는 않다.) 내가 프로젝트에 대해 얘기할 때는 내 의견의 불완전함을 찾기 위해서일 때가 꽤 있다.

     이거는 이런 기성품을 쓰면 쉬운데, "그래도 못생겼어."
     이 글은 이렇게 쓰면 의미가 더 명확해지는데, "읽는 사람은 재수없다 그럴걸."
     여기 청소하기 힘들지 않을까, "건축주분도 이해하실 거야."
     이 회로 또 나누면 스위치가 너무 많아지는데, "그래도 불편해."
     이거 그럴싸해 보이지, "어디서 많이 본 것 같은데."

호숫가의 집
A house by a lake

그럼 나는 예쁘고 쉬운 방법을 찾고, 명확하되 사려 깊은 말투를 고르고, 일 년에 한 번은 빗자루를 댈 수 있는 여건을 만들고, 다른 스위치를 없애는 방법을 강구하고, 그리던 것을 쓰레기통에 버린다. 그렇게 우리의 건축은 이상에 조금씩 다가간다.

     남소장은 말을 투박하게 하는 편이라 좀 알아듣기는 힘들지만 그녀가 하는 말에는 뭐랄까 기똥찬 구석이 있다. 우리가 설계했고 정말 좋아했지만 지어지지 않은 집이 있다. 영화와 책을 좋아하는 한 남자를 위한 집이었다. 의뢰인은 '침실을 만들어줘도 아마 영화 보다가 거실에서 잘 것이니 하나의 큰 거실만 있으면 된다'는 소망을 내비쳤다. 집이 거실 하나만으로 이루어지긴 어려운데 어떡하나 고민하고 있었는데, 남소장이 도넛 비슷한 것을 그리더니

     "그…… 이중 외피…… 크…… 뭐 그런 거 있잖아."
     "이중 외피? 렌조 피아노가 많이 하는 그런 거? 시골집에 뭔 놈의 이중 외피."
     "아니, 그런 거 말고! 그으으으흐…… 뭐 하튼, 그런 거 있잖아."

The Taboo of Balance

Coming to the realization that what you believed to be true turns out to be wrong is a crucial part of making progress. For me, when I express an opinion and hear counterarguments, which I end up accepting, I feel rewarded for that communication. (Even though it feels like that's not always the case for Nam.) Whenever I talk about a project, it is often to find imperfections in what I am saying.

This would be easy if I use a ready-made product like this.
"But it's still ugly."
This writing would make more sense if written like this.
"But readers might think we're jerks."
Wouldn't it be hard to clean this area?
"I'm sure the client would understand."
If I split this circuit again, there may be too many switches.
"That's still inconvenient."
This looks pretty cool.
"That looks kind of familiar."

Then I look for a pretty and easy way, choose a clear but considerate tone of voice, find a way to use a broom once a year, find a way to get rid of some switches, and throw the drawing away into the trash. Like this, our architecture gets close to the ideal, little by little.

Nam tends to speak rather crudely, which makes the conversation a bit difficult but there's always something to what she says. There was a house which we designed and liked very much but never got built. It was a home for a man who loves movies and books. The client expressed a wish, saying, "Even if you create a bedroom, I'll probably end up sleeping in the living room while watching movies. So, I just need one big living room." Not sure if it would be okay to design a house as a living room alone, I was thinking how to work it out. But then Nam drew something like a donut.

"That... double skin... that... you know, something like that."
"Double skin? What Renzo Piano is known for? But who does a double-skin facade for a country home?"
"No, not like that! I meant... anyway, you know what I'm saying."

남소장은 연신 ㅁ자 모양의 도나쓰만 다시 그리고 있다. 이중 외피라……. 건물을 감싸기 위한 건물. 방을 감싸기 위한 방. 집 속의 집. 뭐 이런 거? 하면서 대략적인 평면을 보여 주니,

"그래! 이거 말이야. 이거! 내가 말한 게 바로 이런 거야."

그려 놓고 보니 공간의 바깥을 공간으로 한 겹 더 감싸고 있는 상태를 말한 것 같다. 이중 외피라는 특정 건축 용어를 쓰지 않았더라면 더 이해하기 쉬웠겠지만, 어찌됐건 좋은 아이디어는 맞았다. 우리가 설계한 집 중에 지어지지 않은 집이 세 채가 있는데, 이 집은 그 중에서 가장 아쉬운 집이 되었다.

수롱리
단독주택
House P

우리 사무실은 오랫동안 일이 없었다. 이전 회사를 완전히 그만둔 것은 2014년인데, 우리 사무실의 첫 신축 작업 준공이 2019년이었으니, 일이 없는 기간이 얼마나 길었는지 알 수 있다. 작은 리모델링 작업들을 하고 현상설계에 낙방만 하면서 보내던 그 시절, 나도 좀 피곤해 보고 싶다는 생각을 했었다. 당시에는 한양대학교에서 설계 스튜디오를 가르치고 있었는데 두 사람 다 할 일이 없어서 둘이서 한 스튜디오를 같이 가르쳤다. 일주일 중 가장 중요한 일정이 학교에 가는 이틀이었던 적도 많았고, 거기서라도 건축 이야기를 실컷 할 수 있어서 다행이란 생각도 들었다.

각자 다른 스튜디오를 가르쳤더라면 가계 사정은 좀 더 나았겠지만 그러고 싶지 않았다. 늘 둘이 함께 건축 이야기를 하다 보니, 혼자서 이야기하는 건 불안했다. 나 하나의 의견에 학생들이 반응하는 것을 보는 것보다 우리 두 사람이 서로 다른 이야기를 할 때 반응하는 학생들을 보는 것이 더 흥미로웠다. 둘째 아이가 태어나고, 우리도 일이 조금씩 더 생기면서 더 이상은 같이 한 스튜디오를 가르칠 수 없게 되었다. 이후 몇 년 더 혼자서 스튜디오를 가르쳤지만, 점점 학교에 가는 일이 덜 즐거워졌고 결국 가르치는 일을 그만두었다.

Nam then kept on drawing square-shaped donuts. Double skin... a building to surround a building. A room to surround a room. A house within a house. I then showed a rough drawing, "Something like this?"

"Yes! This is it. This! This is exactly what I was talking about."

Based on the drawing, I think she meant to surround the outer layer with another layer. Perhaps it would've been easier to understand had she not used the specific architectural term "double skin," but anyway, the idea was good. Among the houses we designed, three of them ended up not being built, and this house is the most regrettable one of the three.

수룡리
단독주택
House P

For a long time, we had no work. I completely left the previous job in 2014, and it was not until 2019 that our company's first new construction was completed, which shows how long we survived without work. During the days I spent my days on small remodeling projects and failing at design competitions, I wished I was a little more tired. At the time, I was teaching a design studio at Hanyang University, and since neither of us had anything else to do, we taught together. There were many times when the most important schedule of the week was the two days of teaching, and I felt fortunate that I at least had the teaching job to talk about architecture to my heart's content.

Perhaps we would've been better off financially had taught two different studios, but we didn't want to do that. Since the two of us always talked about architecture together, I didn't feel comfortable being alone. I enjoyed the reactions of students when the two of us said different things rather than when I spoke alone. But after our second child was born and we began to take on other projects, we could no longer teach in the same studio. I taught alone for a few more years, but going to school became less enjoyable and I eventually stopped teaching.

그 집수리
Home
Improvement

우리 회사에 김과 남, 각자의 담당 프로젝트는 없다. 모든 프로젝트를 둘이 함께 생각하고, 두 사람 모두가 동의할 때까지 이야기한다. 사무실에 앉아서 도면을 그리고 현장에 나가는 시간 자체는 내가 더 많지만, 건축적인 의사 결정은 전부 다 같이 한다고 해도 과언이 아니다. 부부가 집에서도 같이 있고 일터에서도 같이 있으면 싸우지 않냐는 질문을 많이 받는데, 사실 그닥 싸우지는 않는다. 서로의 의견 덕분에 앞으로 나아갈 수 있다는 것을 알고 있기에.

그 집수리
Home Improvement

At KimNam Architects, we don't divide projects according to who is in charge. For every single project, we think together and talk about it until the two of us agree. While I may spend more time drawing plans in the office or visiting sites than she does, it wouldn't be an overstatement to say that we make all architectural decisions together. We often get asked how much we quarrel since we're together at home and work, but in reality, we don't that much. Because we know that we are moving forward thanks to each other's opinions.

시간의 의미

2009년, 헤르조그 앤 드뫼롱(Herzog & de Meuron)에서 일을 하게 되어 스위스로 이사를 가게 되었다. 그때만 해도 스위스에 오래 살 것이라 생각하지 않았지만 결국은 4년을 살았다. 사무실이 있던 바젤에서 약 3년 반, <프라콩뒤 주택(Chalet à Pracondu)>이 있던 오뜨-난다(Haute-Nendaz)에서 반 년을 살았다.

주변 사람들은 우리가 스위스에서 살았다고 말하면 '어머, 좋으셨겠어요~'하는 반응을 보이지만, 사실 우리는 이 시기의 삶을 즐거웠다고만 기억하지는 않는다. 20대 후반, 30대 초반이던 우리는 건축가로서 충분히 성장하지 못하고 있는 것은 아닌지 늘 불안해했다. 회사에서 특정한 스터디를 질릴 정도로 반복해야 할 때, 청춘을 갈아 넣은 디자인이 마음에 들지 않을 때, 내가 참여하고 있는 프로젝트가 회사에서 진행하는 가장 멋진 프로젝트가 아니라고 느낄 때 스트레스를 받았다.

비록 반복적인 업무가 많기는 했지만 건물을 설계하면서 스터디에 충분한 시간과 비용이 주어진다는 것이 얼마나 축복받은 일인지 그때는 몰랐다. 훌륭한 컨설턴트들과 협업하는 것, 스터디를 도와줄 장인들이 회사에 상주하는 것이 특별한 일이라는 것도 자각하지 못했다. 누릴 수 있는 것에 집중했더라면 더 행복했을 텐데, 어리석게도 매일 새롭게 얻은 지식이 있었는지 반문하며 조급해했다.

스위스의 높은 물가를 견디는 것, 아이의 의료보험 지원금을 챙기는 일, 도저히 먹을 것이 못되는 스위스 맥주, 일주일에 이틀만 쓸 수 있는 공용 세탁기는 시간이 지나도 적응하기 힘들었다. 자다 깨면 벙글벙글 웃으며 우리에게 기어 오던 귀여운 아들이 없었더라면 일찍이 무너져 버렸을 것이다.

스위스에서 회사를 다니던 시절 콘크리트 샘플을 만들고 있는 김진휴
Jinhyu Kim making concrete samples at a company in Switzerland

회사를 다니던 중 목수인 친구의 <프라콩뒤 주택>과 <르클루 레스토랑(Restaurant Le Clou)>을 계획하게 되면서 이따금 희망 섞인 기분이 들기도 했다. 첫 아이를 낳고 일을 잠시 쉬었던 남호진 소장에게는 이 두 프로젝트가 건축을 놓지 않는 끈이었던 때도 있었다. 안타깝게도 이 일들은 시작하고 멈추기를 반복했다. 계약을 하고

The Meaning of The Time

In 2009, I moved to Switzerland to work at Herzog & de Meuron. Back then, I didn't think I would be in Switzerland for a long time, but I ended up living there for a total of four years. I lived in Basel, where my office was based, for about three and a half years, and in Haute-Nendaz, where *Chalet à Pracondu* is located, for half a year.

When we tell others that we lived in Switzerland, their common reaction is "Oh, lucky you!" But the truth is, it wasn't all roses and sunshine. Being in our late 20s and early 30s, there was always a part of us that felt insecure, feeling we were not growing enough as architects. Whenever I had to repeat a certain study over and over to the point of getting fed up, whenever I didn't like the design I traded my youth for, and whenever I felt that the project I was participating in was not the best project the company was working on, I was stressed out.

Looking back now, I didn't realize how blessed I was to be given sufficient time and money to study while designing even though there was a lot of repetitive work. I didn't realize how blessed I was to be collaborating with great consultants and having craftsmen stationed at the company to help with my studies. Perhaps my life would have been much happier had I focused on my blessings, but I foolishly became impatient, questioning whether I had learned something new every day.

Surviving Switzerland's high cost of living, keeping track of health care subsidies for my child, having no better choices than terrible Swiss beers, and being able to use the shared laundry machine only twice a week were some things I just couldn't get used to. If it weren't for my cute son, who would crawl up to me with a big smile on his face every morning, I would have given up a long time ago.

Perhaps brainstorming to design *Chalet à Pracondu* and *Restaurant Le Clou* for a friend of mine who was a carpenter was a streak of hope during my days of working for a company. The two projects were also what brought Nam back to architecture after her maternity leave. Unfortunately, however, the projects were paused a few times. Rather than a project that proceeds steadily as planned after a contract, these projects were more like an intermittent conversation like, "We'll have to build a house like this

목표를 향해 꾸준히 진행하는 프로젝트라기보다는 '언젠가 이런 집을 지어야 할 텐데'에 관한 간헐적 대화 같았다. 그런데 스트레스 끝에 퇴사를 결심한 그 시점, 우연하게도 스위스의 주택 관련 법이 바뀌게 되었고 의뢰인은 당장 집을 지어야만 하는 상황이 되었다.

우리는 곧 바젤 살림을 정리했다. 남호진 소장은 런던에 있는 친구네 집을 고치러 몇 주간 스위스와 영국을 오가게 되었고, 나는 옷가방과 컴퓨터를 들고 프라콩뒤와 르클루가 있는 오뜨-난다라는 마을로 이사를 했다.

오뜨-난다는 바젤에서 3시간 정도 떨어진 알프스의 작은 마을이다. 건축 열심히 한다는 전세계 젊은이들이 득실대던 바젤과는 달리 이곳은 조용하고 평화로웠다. 프라콩뒤 길(Route de Pracondu)과 르클루(Le Clou) 언덕은 걸어서 5분 정도 떨어진 곳에 있었는데, 이곳을 왔다 갔다 하다 보면 풀 뜯는 소들을 지나치기도 하고 운이 좋은 날은 노루 가족을 만났다. 도시에만 살았던 나는 산에 땅거미가 올라오고, 달그림자가 지는 것을 이곳에서 처음 봤다.

산 속의 삶
Life in the mountain

이 조용한 마을에서는 딱히 할 일이 없었으므로 현장에 가지 않을 때는 앉아서 도면을 그리면 되었다. 나는 불어를 못해서 시공자에게 의사를 전달하기 가장 좋은 방법은 도면을 그려서 보여주는 것이기도 했다. 꼭 말로 할 이야기가 있으면 전날 밤에 불어로 번역해 두었다가 도면이나 사진 위에 뽑아서 들고 가곤 했다. 그리고 목수인 의뢰인이 아주 두꺼운 철물 카탈로그를 몇 권 주었는데 그 속에서 필요한 것들을 찾아보는 것은 꽤 재미있었다. 개스킷, 경첩, 호차 따위를 찾아서 필요한 상세에 그려 넣다가 내가 건축 상세를 좋아한다는 것을 처음 알게 되었다.

The Taboo of Balance

someday." But around the same period that I decided to quit work, it so happened that the housing laws in Switzerland changed and my client had no choice but to build his house right away.

We soon packed our belongings in Basel. Nam traveled back and forth between Switzerland and England for a few weeks to renovate a friend's house in London while I moved to a village called Haute-Nendaz, where Pracondu and Le Clou were located, taking with me just a suitcase of clothes and a computer.

Haute-Nendaz was a small village in the Alps, about three hours away from Basel. This place was quiet and peaceful, unlike Basel that was crowded with young people from all over the world who were passionate about architecture. Route de Pracondu and Le Clou were about a five-minute walk away, and on my way to and fro, I would pass grazing cows and, on a lucky day, a family of roe deer would greet me. As someone who only lived in the city, it was my first time seeing mountains during dusk and the shadow of the moon setting.

산 속의 삶
Life in the mountain

Since there wasn't much to do in this quiet village, I spent my time drawing when I wasn't making site visits. Because I didn't speak French, the best way to communicate to the contractor was through the drawings. For an important message, I would translate it into French the night before and take it with me along with the drawing or photo. The client, who was a carpenter, gave me some hardware catalogs that were very thick, and it was quite fun to look for the things I needed in them. While drawing the details of what I needed, such as gaskets, hinges, door rollers, and so forth, I discovered my interest in architectural details.

프라콩뒤 주택,
현장
Chalet à
Pracondu,
construction

프라콩뒤
주택,
좀
옮겨주시죠
Chalet à
Pracondu,
Could you
please move it?

신축인 <프라콩뒤 주택>과 달리 <르클루 레스토랑>은 리모델링이었고, <프라콩뒤 주택>의 철근콘크리트 공사가 진행될 시점에는 이미 콘크리트 공사와 목구조 공사까지 끝나 있었다. 계속해서 마음을 바꾸는 의뢰인 때문에 <르클루>는 용도가 없는 실내 공간으로 비워져 있었다. 바젤에 살던 의뢰인은 몇 주에 한 번씩 산에 올라오곤 했는데, 그럴 때면 <르클루>에서 뜯어낸 고재들을 같이 정리하거나, 테이블 톱을 가져다 두고 <프라콩뒤 주택>에 설치할 창문의 목업(mock-up)을 만들면서 시간을 보냈다. 의뢰인은 미국에서 유학씩이나 한 놈이 이런 데서 몸 쓰는 일을 하면 되겠냐는 투로 놀려댔지만, 나는 그 일들이 싫을 리가 없었다.

르클루 레스토랑,
창문 목업
Restaurant
Le Clou,
window mockup

르클루 레스토랑,
기존 집에서
떼어낸 마루널
Restaurant
Le Clou,
floorboards
removed from the
existing house

Unlike *Chalet à Pracondu*, which was a new construction, *Le Clou* was a remodeling of an old residential space into a restaurant. By the time the reinforced concrete construction of Pracondu began, the concrete work and wooden construction of *Le Clou* had already been completed. But because the client kept changing his mind, *Le Clou* was left empty without a designated use. The client who lived in Basel would visit once every few weeks, and we would spend our time together sorting through the old materials torn down from *Le Clou* or take a table saw and make mock-ups of windows to be installed in *Pracondu*. The client would tease me, saying what had been an intelligent person in the U.S. was doing wasting my experience like this, but there was no reason for me to dislike what I was doing.

르클루 레스토랑
Restaurant
Le Clou

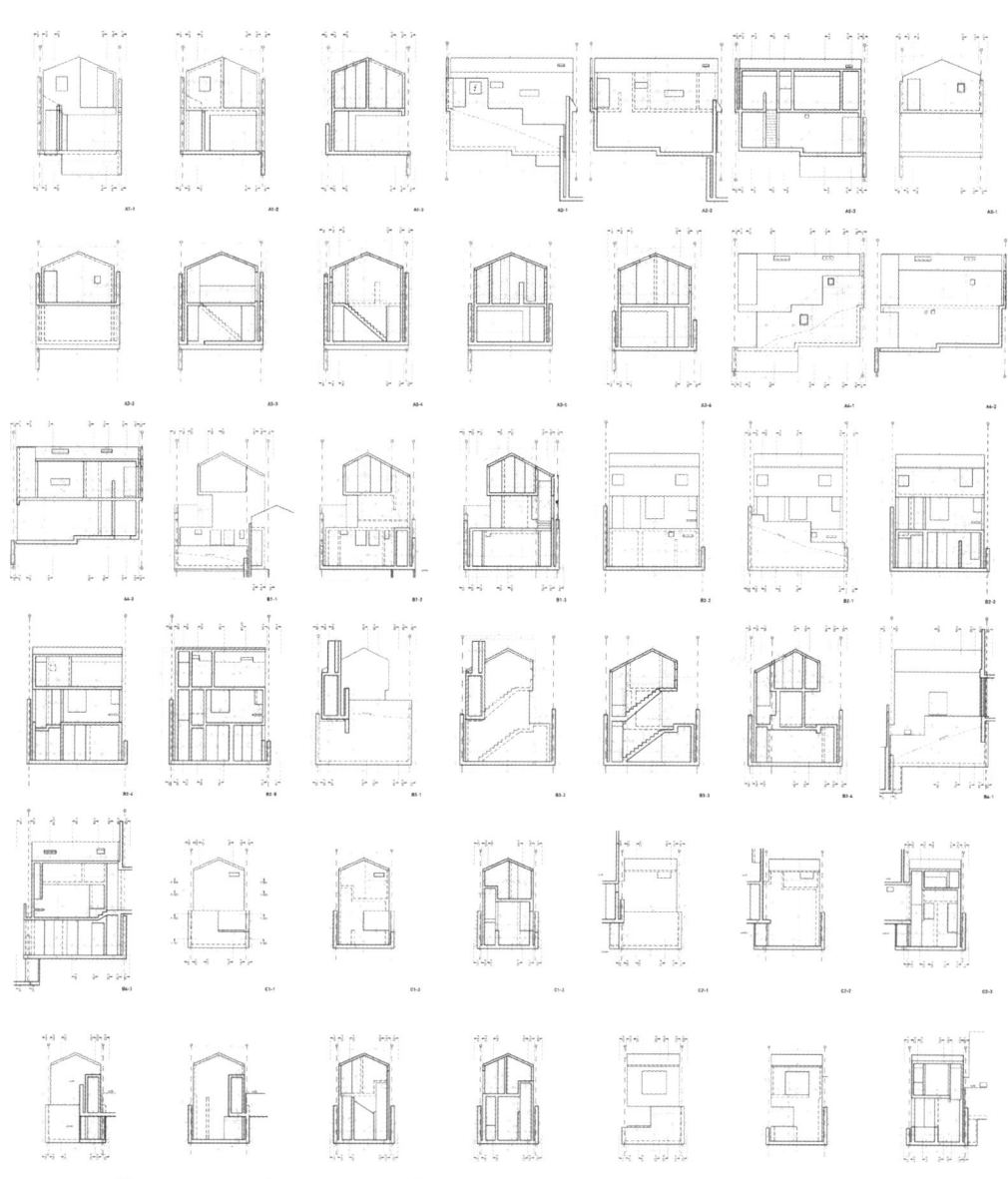

프라콩뒤 주택,
거푸집 입면도
Chalet à
Pracondu,
formwork
drawing

착공한 지 9년이 지난 <프라콩뒤 주택>과 <르클루>는 아직도 지어지고 있다. 심지어 지난 달에는 허가도면과 창의 위치가 맞지 않는다는 이유로 코뮌에서 공사를 중지시켰으니 급하게 입면도를 업데이트해달라는 연락까지 받았다. 착공한 지 9년이 지났는데 아직도 일정이 급하다는 감정을 느낄 수가 있긴 한 건가 생각하면서 10년 전에 그린 도면을 열어 업데이트했다. 깜깜한 밤, 그 시절 산속에서 그랬던 것처럼 모두 잠든 시간 식탁에 앉아 도면을 그리면서 우리에게 이 집들의 의미, 스위스에서 살았던 시간의 의미가 무엇인지 생각해 보았다. 우리가 서두른다고 해서 일이 빨리 이루어지지 않는다는 것, 어차피 마음먹은 대로 되는 일은 별로 없다는 것, 스스로를 괴롭히기보다는 좋은 것을 만들겠다는 마음에 집중하는 편이 더 행복하다는 것을 알려준 것 같다. 그 정도면 좋았었다고 해야 할지도.

르클루 레스토랑
Restaurant Le Clou

It's been nine years since the construction first launched, and *Pracondu* and *Le Clou* are still under progress. In fact, I received a call last month, urgently asking for a revision on the elevation drawings because the location of windows did not match the approved drawings and the commune had paused the project for that reason. Not quite sure if the word "urgent" is appropriate for a project that's been dragging on for nine years, I pulled out the drawing from 10 years ago. Late at night, just like in the mountains back in the day, I sat alone in front of the dining table while others were asleep and worked on the drawing, thinking about what these houses and the time we spent in Switzerland meant to us. I think we learned that rushing doesn't speed things up, anything rarely goes as planned, and life is happier when we focus on the heart to make good things rather than on being so uptight. That said, guess it was good.

그러기로 하는 이유

"오우, 너무나 멋진 드로잉이야. 이거 그린 학생은 당장 액자에 넣어서 할머니께 보내드리도록."

이거 칭찬인가? 잘했다는 건가? 그럴리가 없지⋯⋯ 건축 전공자들끼리 논하는 수업 수준에 전혀 맞지 않는다는 의미였다. 입학한 첫 학기에 듣게 된 이 수업은 피터 아이젠만(Peter Eisenman)의 '시각 연구 입문: 보는 것, 읽는 것, 그리는 것(Intro to Visual Studies:Seeing, Reading, Drawing)'이었는데, 한 학기동안 배우는 내용은 아이젠만의 건축의 자치성과 그 비평에 대한 것이었다.

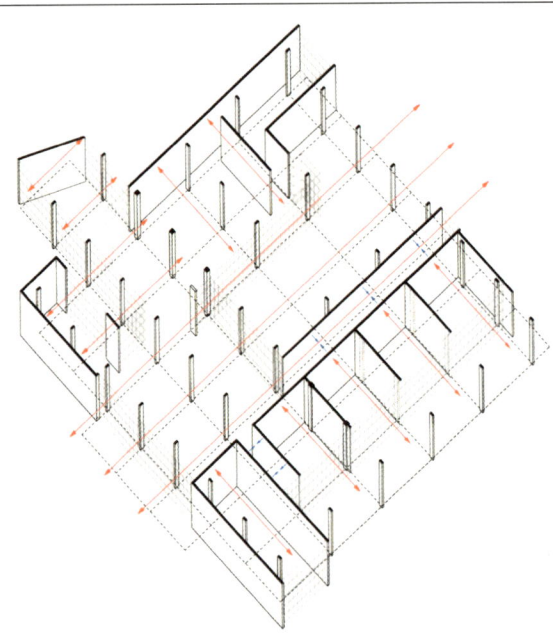

아이젠만 수업 다이어그램, 테라니의 쌍텔리아의 유치원
Diagram of Eisenman's class, Giuseppe Terragni's Sant'Elia Nursery School, ⓒHojin Nam

건축 이론에 관해서는 문외한에 가까웠던 나는 학기가 시작된 지 몇 주가 지나도록 수업을 따라가지 못한 채 헤맸다. 원래 유학생은 수업 하나 정도는 망치는 것이 정상일지도 모른다는 편리한 합리화에 다다를 무렵, 한 학년 위의 어느 남학생이 "이 학교에 다녔다는 사실을 가장 의미 있게 만드는 건 아이젠만 수업일 걸"이라 말하는 걸 듣게 되었다. 그 수업을 듣고 난 후에는 건축을 보는 시각에도 변화가 생기고, 설계를 하는 방식도 달라질 것이라고 했다. 그러고 보니 60여 명의 동기 대부분이 이 수업에 유독 목을 매는 것이 보였다.

아이젠만이 학생들에게 건축의 자치성(Autonomy of Architecture) 혹은 내재적 관점을 이해시키기 위해 채택한 건축의 관점은 '형태주의'였다. 매주 하나씩 어떤 건물이 숙제로 제시되면 그 건물의 형태를 읽고 드로잉으로 표현하는 훈련이 계속되었다. 숙제로 나온 건물들은 대체로 17세기 이탈리아에 지어진 교회들이었다.

The Reason for Doing So

"Wow, what an amazing drawing. Why don't you frame it right now and send it to your grandmother."

Was this a compliment? Did he like the drawing? It can't be... What he meant was that the drawing was not up to the standard of architecture majors. One of the classes I took in my first semester was Professor Peter Eisenman's "Intro to Visual Studies: Seeing, Reading, Drawing," which was about Eisenman's autonomy of architecture and its criticism.

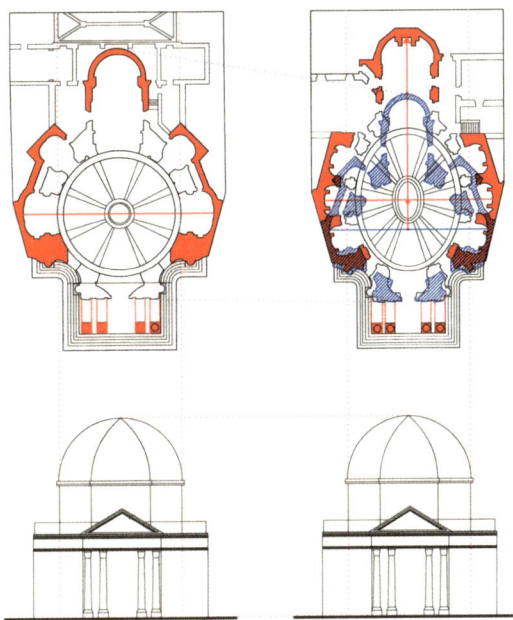

아이젠만 수업 다이어그램, 피아자 델 포폴로의 쌍둥이 교회 Diagram of Eisenman's class, The twin churches of Piazza del Popolo, ⓒHojin Nam

Being near-illiterate when it came to architectural theory, I was lost in class even after a few weeks into the semester. Just when I was about to comfort myself with a convenient rationalization that it's normal for an international student to flunk a class or two, a guy who was a senior to me said, "Eisenman's class is probably the most worthwhile part of coming to this school." What he meant was that Eisenman's class would change my perspective on architecture and also the way I design. As it turned out, almost all sixty or so classmates of mine were particularly desperate over this course.

The architectural perspective which Eisenman adopted to help students understand the autonomy of architecture, or intrinsic perspective, was "formalism." Every week, he would choose a building for us to read its form and express it in drawing. The buildings were mostly churches built in Italy in the 17th century.

누가 의뢰했다거나, 그런 교회를 짓기가 얼마나 어려웠다거나 하는 이야기는 철저히 배제하고 오로지 형태에 담긴 질서와 의도를 추출하고 분석적 도면으로 그려내는 것에만 집중할 것을 요구받았다. 결코 옆길로 새지 않는 글을 쓰는 것처럼.

해는 달랐지만 나와 남소장 둘 다 같은 수업을 들었고, 학교를 졸업할 즈음에는 내재적 건축의 각인이 우리 머릿속 어딘가에 새겨져 있었다. 사용자의 편의, 공사의 경제성, 법적인 제약에 대한 관점뿐만 아니라 건축 분야의 내재적인 관점에서도 건축 작업을 바라보게 되었다. 예를 들면 부분과 전체의 관계, 형태와 경험의 간극, 단순함(또는 복잡함) 같은 것 말이다.

내재적 관점으로 건축을 바라볼 때, 건축은 마치 예술과 유사한 감상의 대상이 된다. 덕분에 건축가들은 사고의 자유를 얻고 다양한 것과 새로운 것을 떠올리는 일에 매진할 수 있다. 공공의 선(善)을 위해서, 기후 위기에 대응하기 위해서, 효율적인 시공을 위해서, 건축주를 위해서가 아니라 그냥 이런 것도 만들어 보기 위해 만들면 되는 것이다. 내재적 관점에서 바라보는 건축의 성패는(만약 그런 것이 있다고 한다면) 얼마나 많은 이들이 이를 흥미롭다고 여기게 되었는지, 더 탐구해 볼 가치가 있다고 느꼈는지에 달려 있다고 할 수 있을 것이다.

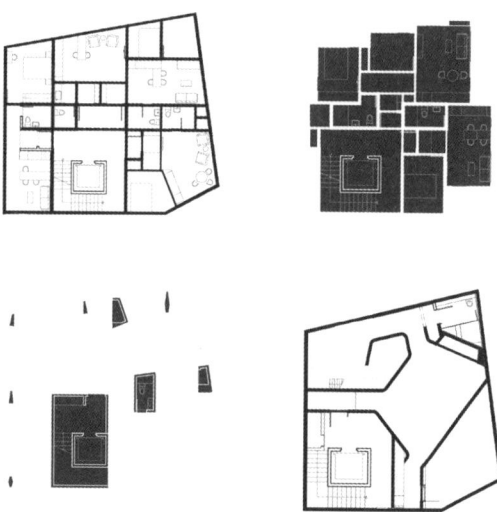

QUAD, 평면 다이어그램
QUAD, Plan diagram

Completely eliminating such factors as who commissioned the work or how difficult it was to build such a church, we were told to focus solely on extracting the order and intention behind the shape for an analytical drawing. As if never going off topic when writing.

Both Nam and I took Eisenman's class, just not at the same time. By the time we were graduating, the imprint of autonomous architecture was engraved somewhere in our minds. We were able to see architectural work not only from the perspective of user convenience, economic feasibility of construction, and legal restrictions, but also from the intrinsic perspective. For example, the relationship between parts and the whole, the gap between the form and experience, and simplicity (or complexity).

From an intrinsic perspective, architecture becomes an object of appreciation, much like art. This allows architects to gain the freedom of thought and can focus on coming up with diverse and new things. Neither for the sake of the public good nor to respond to climate change nor for efficient construction nor for the building owner, but just for the intention of making something. The success or failure of architecture from an intrinsic perspective (if such a thing exists) can be said to depend on the crowd's motivation and whether it is worth further exploration.

QUAD
©Kyoungtae Kim

구청에 <QUAD>의 허가 협의를 하러 간 적이 있었다. 당시 우리 동 담당 주무관은 조만간 출산 휴가를 갈 것으로 보이는 만삭의 임산부였다. 최상부에 있는 다락의 체적과 평균 층고 산정에 대한 나의 설명을 듣던 그녀가 이렇게 물었다.

"꼭 이렇게 그려가면서까지 이대로 하셔야겠어요? 그냥 여기만 평지붕으로 가면 되지 않나?"

"그러면 형태가 훼손되니까요."

"네?"

순간 정적이 흘렀다. 나 스스로도 검열할 수 없을 만큼 내 반응은 반사적이었다. 그녀는 아무 말도 하지 않았지만 마치 오늘 들은 이야기 중에 가장 유치하고 황당하다고 말하는 것 같았다. 어쩐지 부끄러운 기분이 들어서 황급히 변명을 하려고 했다.

"아니, 그게 여기서 지붕이 꺾이면, 그게, 그러니까……"

"아, 알았어요. 암튼 그런 줄 알게요. 법에 안 맞는 것도 아니니까."

주무관도 이 어색한 공기를 얼른 넘겨 버리고 싶었는지도 모른다. 자칫 이 건물에서는 전체를 이루는 부분들의 차이점을 형태, 재료, 평면으로 표현하고자 했고 최상부의 비스듬한 매스에 포함되는 다락의 지붕면에 별도의 요철을 만든다면 '부분 안의 부분'이라는 또 다른 화제가 탄생하므로 맥락을 흐린다는 사실을 설명이라도 할까봐 걱정이 되었을지도 모른다.

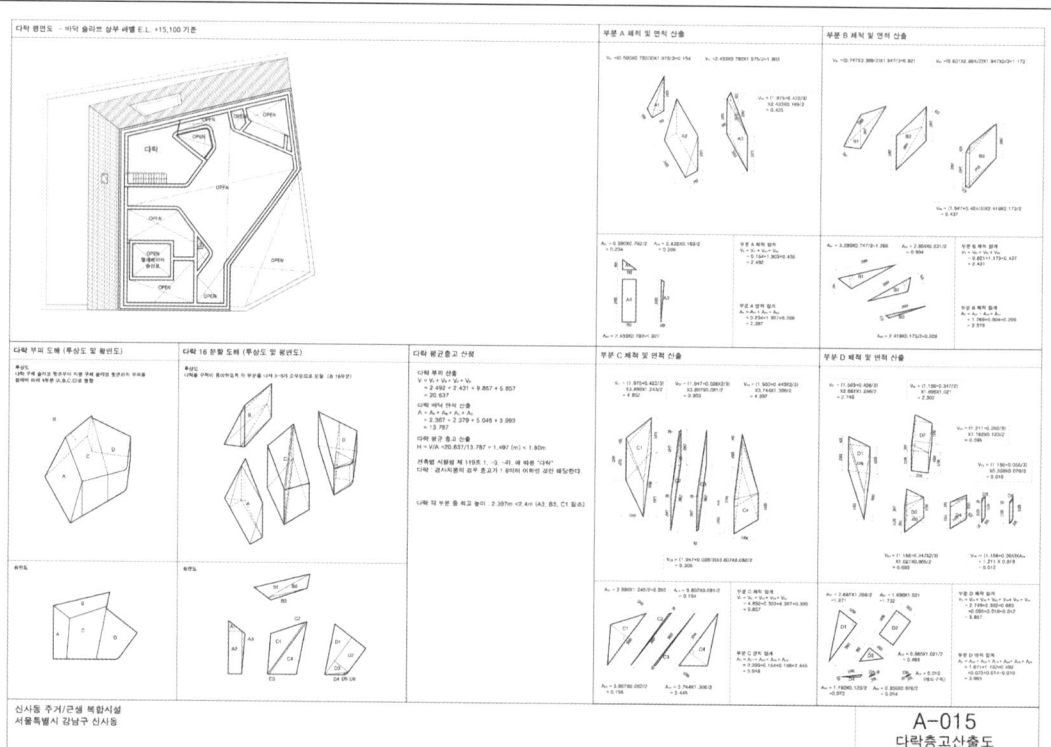

The Taboo of Balance

I once went to a district office to discuss the construction permit for *Quad*. At that time, the surveyor of the neighborhood was a pregnant woman, who looked like she was going to take her maternity leave soon. Hearing my explanation on the volume of the attic at the top and calculation for the average floor height, she asked,

"Do you really have to do it this way with all the drawings? Can't you just have a flat roof here?"

"But that would ruin the form."

"What?"

There was a moment of silence. My reaction was so reflexive that I couldn't even stop myself. She didn't say anything but was probably thinking this was the most childish and ridiculous thing she heard today. Feeling embarrassed for some reason, I quickly tried to make an excuse.

"See, if the roof bends here, um, that means…"

"Ah, okay. Anyway, that's fine. It's not like it's against the law."

Perhaps she also wanted to quickly get over the awkward situation. Maybe she was worried that I would get into the details about the differences between the parts that make up the entire building and how they were expressed with forms, materials, and planes, adding that if we created a separate irregularity on the roof surface of the attic, "a part within a part" would be created, introducing another topic that disturbs the context.

QUAD
ⓒKyoungtae Kim

다시 한번 말하지만 내재적 관점은 쓸모나 비용이나 남을 위하는 일과는 무관한 것으로서 남을 설득하는 일에는 하등 도움이 되지 않는다. 따라서 이 관점을 바탕으로 한 판단에 대해서는 건축주, 허가권자, 부모님 앞에서는 결코 발설하지 않는 편이 좋다. 그럼에도 불구하고 우리끼리 있을 때는, 그러니까 건축을 만드는 일을 하거나, 건축을 읽는 일을 하거나, 건축이 발전하기를 바라는 사람들끼리 있을 때는 가끔 이야기해도 좋을 것 같다. 핑계 대는 습관을 버리기 위해서, 온통 윤리적인 척하지 않기 위해서, 건축의 다양성에 아무런 제약도 만들지 않기 위해서, 진심으로 좋아한다는 사실을 인정하기 위해서.

<이야기, 공예>
부여 백제 금동
대향로 전망대
설계공모 제출안
<Story, Artistry>,
Gilt-bronze
incense burner of
Baekje, Buyeo
design competition
Proposal

I repeat, the intrinsic perspective has nothing to do with practicality, cost, or serving others, and is definitely of no help in persuading others. As such, it is recommended that you keep your judgment based on this perspective to yourself; not to be shared with the client, permitting authority, or your parents. There is one exception: It might be a good idea to talk about it when we're alone, that is, when we are with those who practice architecture, read architecture, or hope for the development of architecture. That way, we would give up the habit of making excuses, stop pretending to be entirely ethical, not make restrictions on the diversity of architecture, and admit how much we really enjoy what we do.

The Taboo of Balance 균형이라는 금기

Lighter than colors

남 이야기

"이제 뭐하지? 이제 뭐하지……?"

유달리 얼굴이 창백한 1학년 후배. 학기말 마감 전날 밤 시간이 남았는데 무엇을 더 하는 게 좋을지 모르겠다며 황망한 표정으로 스튜디오를 떠돌고 있었다. 아니, 지금, 더 할 게 없다고? 시간이 남았는데? 마감 당일 바로 앞 발표자가 발표 중인데도 플로터 앞에서 내 패널이 출력되기를 기다리며 종종거리기 일쑤였던 나로서는 저건 도대체 뭐하는 인간인가 싶었다. 그때의 나는 모형이든 도면이든 끝까지 붙들고 앉아서 아직 시간이 남았다며, 더 할 수 있다며, 잠 한숨 못 자고 달리는 것이 불꽃 같은 열정이라 믿었다. 세상에 없는 것을 만들고 싶었고 모두를 감동시키고 싶었던 시절이었다. (지금 와 생각해보면 걸어 놓고 나면 티도 안 날 것에 집착하느라 정작 중요한 발표 준비만 못했던 것 같다.) 그에 반해 '황제마감'[1]을 꼬박꼬박 해내는 후배에 대해서는 그 허여멀건한 얼굴만큼이나 허약해 빠진 프로젝트를 했음이 틀림없을 것이라 생각했다.

대학원 재학 시절 김진휴의 책상
During graduate school, Jinhyu Kim's desk

---

1 마감인데 세 끼 챙겨 먹고 잠 자고 씻고 발표장에 올 수 있는 사람은 다른 꼬질꼬질한 학생들과 달리 마치 황제 같다고 하여 황제 마감이라고 불렀다.

This is Not My Story

"What do I do next? What do I do next......?"

He was a first-year student and extremely pale. On the night before the end-of-semester project deadline, he would be wandering around the studio, looking lost and not knowing what to do. What, there's nothing more to do?? But there's time left!! As someone who used to anxiously wait in front of the plotter for my panel to be printed even though it's my turn to present next, I couldn't stop wondering what on earth he was doing. Back then, pulling an all-nighter, holding on to it, whether a model or a drawing, was considered passion. Those were the days I wanted to create what didn't exist and wanted to impress everyone. (Now that I think about it, I was too obsessed over the small stuff rather than the important presentation itself.) As for that guy, who was considered "the god of deadlines,[1]" I was certain his project was as flimsy as his haggard face.

---

1 If you never skipped a meal and didn't pull an all-nighter before a deadline, and even managed to shower on the day of the presentation when everyone else is looking all scruffy, you were considered the god of deadlines.

내가 다닌 학교는 모든 학년이 동일하게 월요일과 목요일에 설계 스튜디오가
있었다. 스튜디오가 끝난 목요일 저녁이면 동기나 후배들이 모여 저녁을 먹고
맥주를 마시곤 했었다. 그럴 때 간혹 저녁을 같이 먹던 사람 중 한 명이었던 그 후배와
더욱 가까워지게 된 계기는 내 마지막 스튜디오 마감을 하면서였다. 당시 나는
뉴욕의 설계사무소 SHOP의 어드밴스드 스튜디오(Advanced Studio)를 듣고
있었고, 내 프로젝트로 말하자면 물론 매우 야심 찬 설계였다. 한 주 먼저 마감을
끝낸 후배에게 전화를 걸어 마감 모형을 좀 도와달라고, 에스컬레이터 몇 개만 좀
만들어달라고 했을 때, 그는 몰랐으리라. 이것이 인도 델리에 짓는 국제 공항의
설계였으며, 만들고자 하는 모형에 40개도 넘는 에스컬레이터가 들어간다는 사실을.
공항임에도 불구하고 과감하게 1/100 스케일의 모형을 만들기로 했기에 성인
두 명이 겨우 들만한 꽤 크고 무거운 모형이 될 것이라는 사실을.

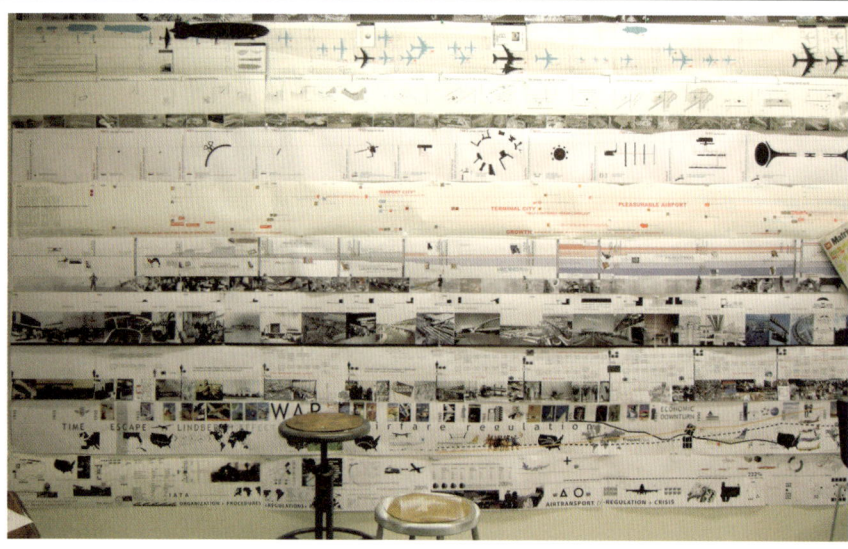

2008년 Shop Advanced 스튜디오 풍경. 공항에 대한 공동 스터디
In 2008, Shop Advanced studio scene. A joint study on airports.

The Taboo of Balance

My school had design studios Mondays and Thursdays for every year. After class on Thursday evenings, we often got together for dinner and some beer. Once in a while, he also stayed for dinner, and it was during my final studio deadline that he and I got closer. At the time, I was taking the Advanced Studio at SHoP Architects in New York, and you could say my project was a very ambitious one. When I called him and asked for help with a few escalator models because he had already finished his project a week earlier, probably little did he know that I was designing an international airport for Delhi, India, that I needed over 40 escalators, and that this was going to be a fairly large and heavy model that could barely be lifted by two adults because I boldly decided to scale it to 1:100 even though it was an airport…

김진휴의
책상
Jinhyu Kim's
desk

본인의 마감을 막 끝낸 후배는 '에스컬레이터 몇 개만' 만들어 주러 3학년 스튜디오에 들렀다가 닷새동안 매일 새벽 서너 시까지 붙들려 대책 없이 일을 키워 놓은 두 명의 선배 (2인 팀 작업이었다.)의 특대형 모형의 많은 부분을 만들어 주는 경험을 하게 되었다. 판을 키운 많은 일들의 결과가 그러하듯이 이 모형은 아무리 밤을 새워도 도무지 끝날 것 같지 않았다. 마감을 위해서는 도면도 그리고 이미지도 만들어야 하는데 그 조화도 무너져 버려 굉장히 스트레스를 많이 받는 상황이었다. 난데없이 남의 마감에 끌려와 잡혀 있는 얘는 얼마나 짜증이 났을까. 내 팀원이었던 데이비드 (David)가 괜찮냐고 물었을 때, 이 후배는 "나는 그래도 지금 배우고 있어."라고 답했다고 한다. 이 황당한 말이 요즘도 가끔 생각이 난다. "그래, 나도 지금 배우고 있는 것이다……"라고 되뇌어 본다.

모형을 돕고 있는 김진휴
Jinhyu Kim, who is assistinng with make the model.

약간 사회성이 떨어지는 후배는 어디를 가도 인싸는 절대로 못 되지만 그렇다고 또 나쁜 평판 같은 건 남기지 않는 모범생이었고 졸업할 즈음엔 이미 유럽의 유명 회사에 취직도 되신 몸이었다. 타고 다니던 차에 이삿짐을 가득 실어 해외 이사를 하고 혼자 먼저 가서 살 집을 구해둔 후, 아내와 생후 두 달 된 아이를 맞이했다. 그렇게 미국을 떠났던 후배는 많은 이들이 부러워한 그 회사를 3년 조금 넘게 다니더니, 굳이 그만두겠다고 했다. "이제 조금 있으면 좋은 기회로 한국에 파견을 나간다더니 조금만 더 참아보지 그래~"라고 말하고도 싶었지만 그러지 않았던 것은 그가 그럴 때는 그럴만한 일이었을 것이라 생각했기 때문이다.

So it turns out, the god of deadlines, who made a casual visit to a third-year student's studio to help make "a few escalators," ended up spending the next five days, staying up until three or four in the morning, making many parts of the extra-large model that the two seniors (it was a two-person teamwork) of his started without any planning. As is often the case with raised stakes, the project seemed like it was never going to finish. In order to complete it, we needed to draw plans and create images, but the harmony was already lost and we were all extremely stressed out. Come to think of it, how frustrating must he have been, having to be dragged into someone else's deadline out of nowhere. When my teammate David asked if he was okay, he said "I'm at least learning right now." His absurd response comes to mind from time to time. "Yes, I am learning something right now," I tell myself.

남호진
졸업학기
마감 모형
The final project model for Hojin Nam's graduation semester

Although far from being popular because he wasn't very social, that junior of mine was one of the top students who never left a bad reputation, and he had already landed a great job at a famous European company toward graduation. He loaded his car with belongings and moved abroad. After finding a place to live, he then welcomed his wife and their two-month-old child. Having to work at that dream company for a little over three years, he insisted on quitting. I felt the urge to persuade him, saying, "Why don't you hang on just a little more since you said you will be able to relocate to Korea soon for a better position?" But I held back because he must have had his reasons.

그 후배가 조그만 회사를 같이 시작하자고 했을 때 나의 마음은 다소 방관자에 가까웠다. 아기가 아직 어려 내 손이 많이 가던 때이기도 했고, 미국에서 공부하고 일을 했던 야심가의 시간 이후로는 스스로 좀 지쳐있었던 것 같기도 하다. 해외 대형 설계회사에서 경험한 실무가 새로 시작한 작은 회사의 일들과는 좀 거리가 있어 무얼 어떡해야 할지 막막했던 것도 사실이다. 어쨌거나 우리는 3년 정도 일이 없어 고생했다. 드디어 첫 건물을 짓고, 그 건물이 건축 잡지에 소개되고 하는 일들은 더없이 천천히 일어났다. 해맑았던 후배가 점점 더 작아지는 모습을 보일 때쯤, 회사로 건축설계를 의뢰하는 전화들이 한 통 두 통 걸려 오기 시작했다. 그리고 어느덧 회사는 정신없이 바빠졌다. 나는 클라이언트 미팅을 나가고 중요한 의사결정을 했지만, 후배는 그것들도 물론 다 같이 하고, 사무실에서는 도면과 3d 모형을 만드는 실무자의 역할, 하루에 전화를 70통씩 받는 프로젝트 매니저의 역할, 현장에서 잘못 시공된 부분을 찾아내야 하는 감리자의 역할, 회사의 경리 및 IT 담당자 역할까지 겸하고 있었다. (게다가 집에 돌아가면 그렇게 부인이 아들 수학을 가르치라고 한다고 했다.) 꿈꾸던 소년은 현장 작업자의 노고와 클라이언트의 불만, 직원들의 눈치 등을 살필 수밖에 없는 현실의 소장이 되어 가고 있었다.

김진휴의 뒷모습
Jinhyu Kim's backside

어렸을 때 길을 걷다 보면 건물 간판에 '부부치과'라고 적힌 걸 볼 때가 있었는데 그럴 때면 "엥? 저게 뭐야? 설마 저 치과에 가면 부부가 있다는 거야?"라며 그 표현을 굉장히 어색하고 낯설게 느꼈던 기억이 난다. 최근 어느 자리에서 "건축계에는 왜 그렇게 부부가 많아요?" 질문을 받은 적이 있는데 아마도 삶과 일을 분리하지 않는 태도 때문인 것 같다고 답했다. 질문을 던졌던 분 또한 아티스트로 창작자였는데, 웃으며 "우리도 삶과 일 분리 안 하는데, 그래도 같이 작업은 안 하는데~"라고 했다.

동창들을 만난 자리에서 "남편이랑 일하면 싸우지 않아? 어떻게 그렇게 해?"라는 질문을 받았을 때 "글쎄. 우린 처음부터 지금까지 계속 아침, 점심, 저녁을 같이 먹어."라고 대답하여 친구들을 경악하게 한 적이 있다. 그저 하루 세끼를 함께 하는 것이 다가 아니다. 잠을 자면서도 일을 하는 건지 밤에 눈을 감으며 "근데 말이야,

The Taboo of Balance

When that junior of mine asked me to start a small company together, I was more like a bystander. I had a young baby who still needed my care, and I think I was a bit drained after my ambitious days, studying and working in the U.S. On top of that, I wasn't too sure what to do since my practical experience at a large international company did not seem relevant to what needed to be done at a small, new company. Anyhow, we spent about three tough years without work. Time passed too slow until our first building was built and featured in an architecture magazine. Around the time when that bright junior of mine began to lose a smile on his face, we started receiving phone calls, one after another, inquiring about architectural designs. And before we knew it, we were swamped with work. While we both attended client meetings and made important decisions, the junior of mine, on top of that, had to create drawings and 3D models, answer 70 phone calls a day, make site visits for errors, and even handle the company's accounting and IT duties. (When at home, I heard, his wife pressures him to teach his son math.) The boy who was a dreamer was now living the reality of a representative with no choice but to endure the demands of site workers, complaints of clients, and mood swings of his employees.

When working together

I recall walking down the street when I was young and seeing a dental clinic with an "a husband-and-wife team" business sign, which made me think, "What? Husband and wife work together?" To me, that was very awkward and unusual. Recently, I was asked, "Why are there so many couples working together in the architecture field?" My answer was probably because of not separating personal life and work.

The person who asked the question was also a creator, being an artist, and their reaction was, "We don't separate our personal lives and work, too, but we don't work together." The time I met my friends, they asked me, "How do you put up with working with your husband?" "Well," I would answer much to their surprise, "We've been having every meal together from the beginning." Here, I'm not just talking about eating three meals a day. Not sure

그 창틀 말이야." 하던 그는 아침에 눈을 뜨며 "근데 말이야, 그 코너 있잖아."라고 한다.
    점심을 먹을 때도 물론 그 '근데 말이야'는 계속된다. 하물며 그 이야기는 아주 아무렇지 않은 듯 건축의 역사니, 우리가 나아가야 할 방향이니 아주 먼 곳까지 가 버린다. 어떨 땐 딴생각을 하다 들켜서 그에게 원망을 듣기도 하고, 어떨 땐 공감 능력 떨어지는 그에게 내가 빈정이 상해버려서 "어차피 당신 회사니까 알아서 해. 회사에서 내 이름 빼줘."라고 할 때도 있긴 하지만……
    나와 매우 다른 후배 김소장과 내가 사는 모습이다.

whether he's working in his sleep, his last words before falling asleep would be something like, "But, that window frame…" Then opening his eyes in the morning, he would say, "But, you know, that corner over there."

Of course, the "but" continues even during lunch. Sometimes, the conversation goes very far, very smoothly, to the history of architecture or the direction in which we should take. Sometimes, my wandering thoughts get caught and crushes his soul, or other days, I would give him a cold-eye stare for lacking empathy with me and say, "It's your company after all. Just take my name off and do what you want."

This is how I live with Kim, a junior of mine who is simply a different existence from another planet.

호숫가의 집 A house by a lake

두 편의 에세이

오래된 오두막
Old cabin

진지한 화답(2021 젊은 건축가상 지원 에세이)

1.

어느 날 목수 일을 하는 친구가 집을 짓고 싶다고 했다. 그의 아내가 살고 있는 산 속에는 쓰임을 잃고 방치된 오두막들이 많이 있는데, 그 집들이 사라지는 것이 안타깝다고 했다. 이름 모를 목수에 의해 백여 년 전 만들어진 그 오두막들을 사용해서 새 집을 짓고 싶다는 그의 바람이 우리 사무소의 시작이었다.

충분한 실무 경험을 쌓은 것은 아니었지만, 다니던 회사를 그만두고 사이트 옆으로 이사를 갔다. (건축주 가족 이외에는) 아는 사람이 한 명도 살지 않는 산 속이었다. 허가도면, 구조도면, 그리고 몇 장의 간소한 도면만으로 집을 짓기 시작했다.

늘 쓰던 도면 양식이나 복사해서 쓸만한 상세 같은 것은 없었다. 누군가 "이 부분은 어떻게 할 거냐" 물어 오면 그제서야 빈 화면에 선을 하나씩 그려가며 도면을 만들어갔다. 다행히 우리는 시간이 많았다. 이 집은 나무(옛 것)와 콘크리트(새 것)의 조용한 대화 같았고, 창과 문과 지붕과 단열재 같은 것들이 어떻게 자리할 것인지 고민하면서 도면을 그려 나갔다.

The Taboo of Balance

Two Essays

<u>Sincere Response (Application Essay for the KYAA 2021)</u>
1.
One day, a friend who was a carpenter said he wanted to build a house. He said there were many empty and abandoned cabins in the mountains where his wife was living, and it was a pity seeing them disappear. His wish to build a new house using the old cabins built by an unknown carpenter about a hundred years ago became the beginning of our office.

Although I could've definitely used more practical experience, I quit my job and moved next to the site. It was in the mountains where I was a complete stranger (except for the owner's family). With only a permit drawings, structural drawings, and a few simple drawings, and a few simple drawings in my hand, I began to build a house.

There was no drawing format that I had always used or any details that could be copied and used. When someone asked me, "What are you going to do with this part?", I would then start drawing lines on a blank screen. Luckily, we had plenty of time. This house was like a quiet conversation between wood (old) and concrete (new), and I thought of how to place things like the windows, doors, roof, and insulation as I drew the plan.

철물 백과를 펴 놓고 필요한 부분을 그렸다. 아는 만큼만 그릴 수 있기에, 대상들의 관계에 대해 필요한 만큼의 정의를 만들어 간다는 생각으로 상세를 만들었다.

서울에 온 뒤로도 그러한 습성은 유지되었다. 조금 더 솔직해지자면, 라이브러리를 구축해 두고 효율적으로 완성도 있는 도면집을 만드는 프로세스에 대해 무지했기 때문에 발생한 일이었다. 그러나 건축의 요소들이 만나는 상황에 대해 통상적인 해법에 의존하지 않을 때 고유한 상세가 만들어지기도 한다. 우리는 그 상황이 우리의 건축이 멈추어 있지 않게 해 준다고 생각한다.

오래된 오두막의 나무 표면
The wooden surface of the old cabin

새집의 콘크리트 표면
The concrete surface of the new house

The Taboo of Balance

For hardware parts, I drew what I needed with the hardware encyclopedia open. Since I can only draw as much as I know, I filled in the details with the idea of giving as much definition as necessary to the relationships between objects.

    This habit of mine continued even after I relocated to Seoul. Frankly speaking, this is what happens when you are ignorant of the process of having a library that helps effectively create a near-perfect collection of drawings. The unique details, however, sometimes come together when architectural elements meet and do not rely on conventional solutions. We believe that it is those situations that keep our architecture continuing on.

프라콩뒤 주택
Chalet at Pracondu

2.
우리는 건축에 대한 극단적 다원주의자이다.
우리는 안심하고 지속적으로 좋아해도 될만한 것을 찾지 못했다. 이전에 싫다고 생각했던 것을 나중에 좋아하게 된 적도 있고, 어제 믿고 있던 사실이 오늘 틀려 보였던 적도 있다. 가치관이 없다는 것이 아니다. 어떤 것에 대해 오늘 좋지 않다고 생각했다고 해서 내일도 싫어할 것이라고 단정짓지 않는다는 것이다.
건축에서 일관성은 추구해야 할 가치인가.
자연재료의 빛깔을 그대로 드러내는 것은 페인트로 색을 만드는 것보다 고결한 일인가.
완전함은 좋은가. 불완전함은 좋은가.
치우침은 경계해야 할 대상인가. 균형은 경계해야 할 대상인가.

오래된 집, 불완전한 것
An old house, imperfect

3.
부재(不在)는 생산성을 담보하지 않는다. 라이브러리의 부재는 비효율적이고, 진리의 부재는 피곤하다. 그러나 이러한 부재는 우리가 프로젝트를 통해 만나게 되는 여러 문제에 진지하게 화답하는 방식이 되어 왔다.
이전 프로젝트에서 성공적으로 구현되었던 아이디어라고 하여 그것이 다음 프로젝트에서도 기준이 될 수는 없다. 어제 그린 평면이 많은 장점을 가지고 있다고 해서 그 평면이 유일한 정답이라고 단정하지 못한다. 그렇게 우리는 계속 의심하고, 계속해서 고쳐 그린다. 아직 보지 못한 무엇 중에 더 나은 것이 있을지도 모르기 때문에. 더 좋아할 만한 무언가를 만나고 싶다는 마음으로.

The Taboo of Balance

2.
We are extreme pluralists when it comes to architecture.
    We haven't found something we will continue to like in relief. There are things we disliked before but ended up liking later, and there have been times when what I believed to be right yesterday seemed wrong today. It's not that we have no values. It's simply that we never say never.
    Is consistency a value to pursue in architecture?
    Are the colors of natural materials considered more noble than created colors using paint?
    Is perfection good? Is imperfection good?
    Is bias something to be wary of? Should we be wary of balance?

색으로 가득 찬 공간
A space filled with color

3.
Absence does not guarantee productivity. The absence of a library means inefficiency, and the absence of truth means exhaustion. But such absence has become a way of responding seriously to the various problems we encounter through projects.
    Just because an idea was successfully implemented in a previous project does not mean it can be the standard for the next. Just because the drawing from yesterday had many advantages doesn't necessarily make the drawing the only solution. Like this, we continue to doubt and resketch, making one correction after another. For something better out there yet to be seen. Longing to meet something more attractive.

김남건축 | KimNam Architects

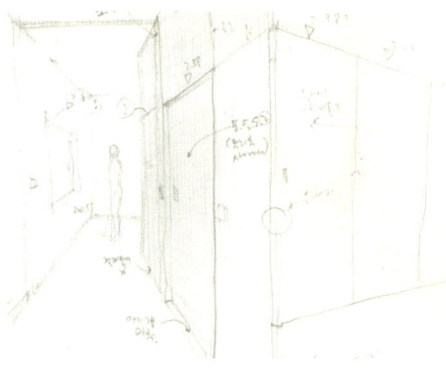

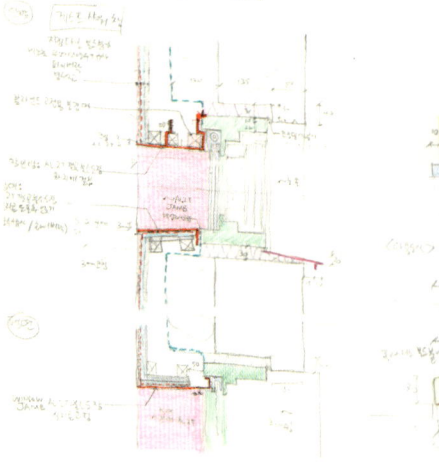

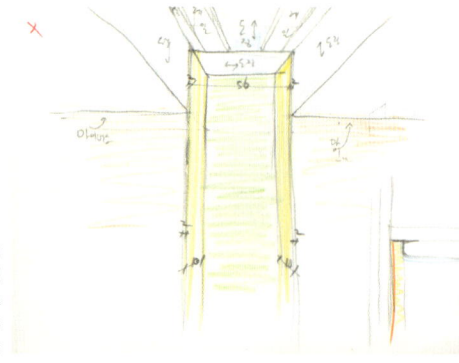

고유한 디테일이
만들어지는 것
The creation
of unique
details

The Taboo of Balance | 균형이라는 금기

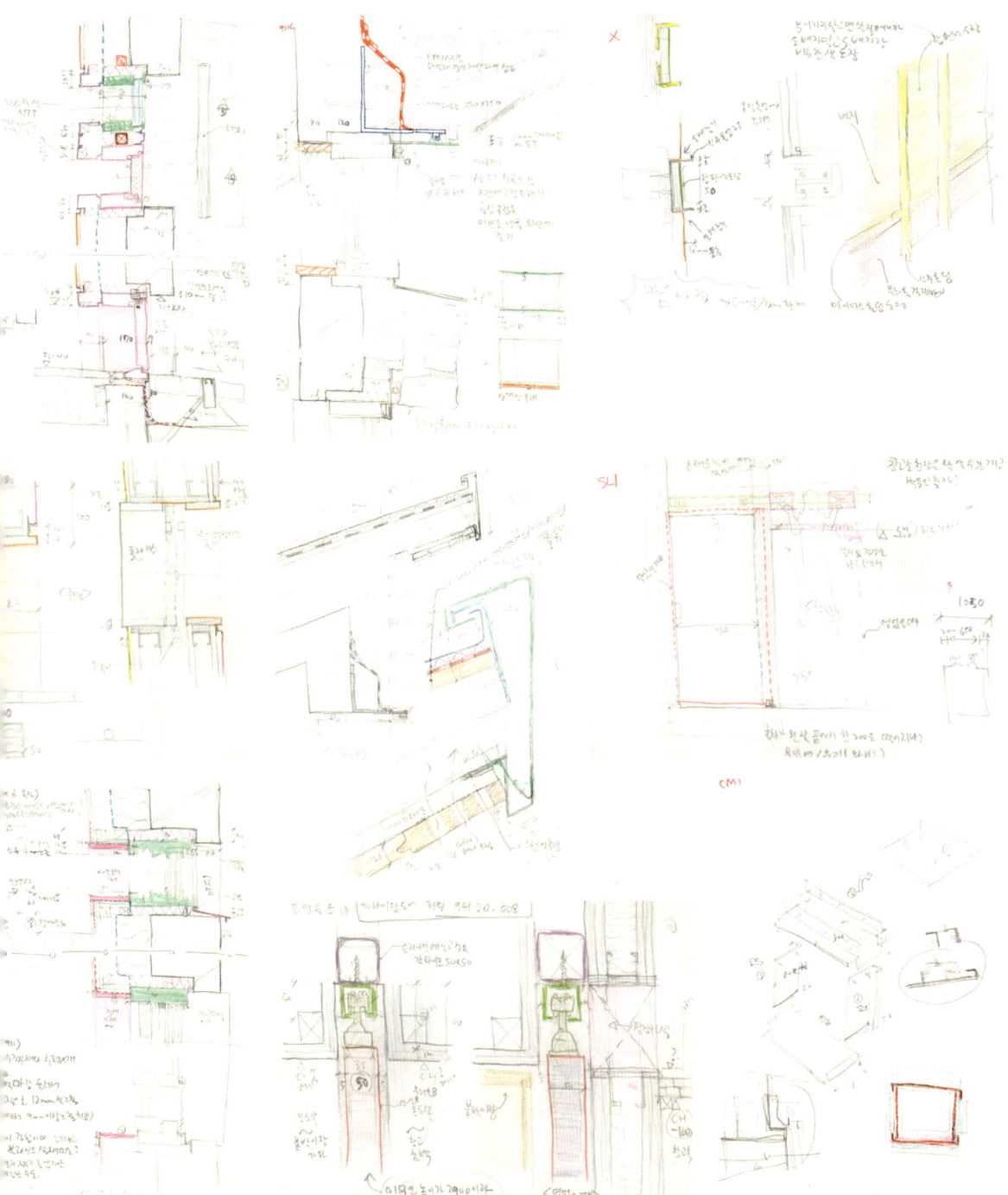

아름다움에 이르는 길(2023 젊은 건축가상 지원 에세이)

우리에게 설계를 맡기기 위해 찾아온 사람들은 이런저런 이야기를 들려준다. 이야기는 보통 대지면적이나 공사비 같은 것으로 시작하지만, 화제는 금세 본인들의 인생에 대한 것으로 넘어간다. 용돈을 받고 잡초를 뽑다가 지쳐버린 아들과 딸에 대한 이야기, 언제부터인가 거를 수 없는 일과가 된 반신욕에 대한 이야기를 들으며 우리는 그들의 삶에 대해 조금씩 이해하게 된다.

먼저 말 걸어 줄 사람이 없는 설계 공모에서는 우리의 기억, 또는 살면서 만난 사람들의 기억을 바탕으로 건물과 관계를 맺게 될 사람들을 떠올린다. 계단이 너무 큰 난관이었던 어느 뚱보 유치원생, 장마철을 앞두고는 배수구를 전부 열어 보아야 잠을 편히 잔다는 관리인, 유모차를 끌고도 갈 수 있는 화장실을 찾아 헤매는 초보 엄마의 입장이 되어 본다.

누군가를 이해하고 공감한 뒤에는, 건물에 담길 그들의 삶을 상상하기가 수월해진다. 가족 구성원 모두가 성인인 집의 분주한 아침 풍경이라든가, 어린 아이 15명을 데리고 계단을 오르내리는 유치원 선생님의 얼굴 표정 같은 것들 말이다. 건물이 말을 하는 것도 아닌데, 건물에는 웃음과 짜증, 불안감과 감사함 같은 마음이 묻어 있는 것 같다.

이 도시의 유치원
Kindergarten of the city

The Path to Beauty (Application Essay for the KYAA 2023)

Clients who come to us for designs are like storytellers. Their stories usually start with the land area or construction costs, but soon enough, they move onto their life stories. The story of their sons and daughters who grew tired after pulling weeds for an allowance or the story of how a lower-body bath became an indispensable daily ritual, helping us to understand them a little more.

In a design competition where we have no such storytellers to guide us through, we rely on our memories – or the memories of those we crossed paths with – to imagine those who will be a part of the building. Like the chubby kindergartener for whom the stairs were too much of a challenge, the janitor who had to check all the drains before the rainy season to sleep comfortably, and a new mother desperately in search of a public bathroom that is stroller accessible.

Once we understand and empathize with someone, it becomes much easier to imagine the life story that will unfold in the building. Say, all-adult family members rushing through a busy morning or the expression on the face of a kindergarten teacher walking up and down the stairs with 15 young children. Building doesn't speak, yet it seems to harbor emotions like laughter, irritation, anxiety, and gratitude.

이 도시의 유치원, 스케치
Kindergarten of the city, Sketches

KimNam Architects

우리의 상상들은 차츰 건물이 갖추어야 할 좋은 부분이 무엇인지 알려준다. 의뢰인의 이야기에 담긴 소망은 이런 부분들에 대한 직접적인 단서가 되기도 한다. 하지만 우리의 생각은 그것에 그치지 않는다. 집 앞 골목길에 드리우게 될 그림자, 서로 다른 속도로 낡아갈 외벽의 재료들, 세대가 바뀐 뒤 고쳐 쓰게 될 집의 모습처럼 더 넓은 범위로 생각이 퍼져 나간다.

우리의 생각이 건물 안 누군가에 대한 것을 넘어 건물을 둘러싼 사회나 건물이 존재할 시간에 대한 것으로 바뀔 때쯤 우리는 건물 설계에 완전히 몰입해 있다. 처음에는 사소했던 우리의 상상은 감당하기 힘들 정도로 복잡하고 어려운 문제들을 낳아 두었다. 포기하면 편한데 결국 만족시키고 싶은 것들, 완공되고 나면 보이지도 않을 부분들을 또다시 고민한다. 간결하기 위해 필요한 것을 생략한 뒤 느꼈던 후회, 더 넓고 집요하게 고민한 끝에 어려운 것을 달성했던 기억이 우리를 계속해서 몰두하게 한다.

이 도시의 유치원
Kindergarten
of the city

The Taboo of Balance

Our imaginations gradually reveal what good aspects the building should embody. The wishes told in a client's story can also serve as direct clues to these aspects, but our thoughts do not stop there. We spread our thoughts to a wider range, such as the shadows that will be cast on the alleys in front of the house, the materials of the exterior walls that will wear out at different rates, and the appearance of the house that will change over generations.

By the time our thoughts shift from those inside the building to the society surrounding the building or the time period in which it will exist, we are completely immersed in the design of the building. Our imaginations that were trivial at first have left behind problems overwhelmingly complex and challenging. It would be easier to just give up, but we cling onto the things we want to satisfy and the parts that won't even be visible once it's completed. The regret we felt after omitting something necessary for the sake of brevity, or the memory of accomplishment after hours and hours of being persistent keeps us going.

우리는 달성하기 어려운 무언가를 이루었을 때, 그렇게 만들어진 대상에서 아름다움을 느끼곤 한다. 좋지만 만들기 어려워서 쉽게 찾아볼 수 없는 것을 마주하면 우리는 고유함, 신선함, 귀함을 떠올린다. 모두가 동의하지는 않겠지만, 우리에게 상대적 희소성은 황금비율과 같은 기하학적 규칙보다 아름다움에 더 가까이 있다. 이때 아름다움은 이 시대의 문명, 이 사회의 경제 구조, 기술적 수준과 깊이 연결되어 있다.

우리가 삶에 대해 상상하는 것을 멈추지 않는 이유는, 만들어 주고 싶은 좋은 것을 찾기 위해서이다. 계속되는 상상은 약속이라도 한 듯 버거운 숙제를 내주고, 우리는 집요하게 그 숙제를 해낸다. 그것이 우리가 생각하는 아름다움에 이르는 길이다.

왼쪽.
스터디 모형
Study models

가운데.
사무실 풍경
In the office

오른쪽.
QUAD
5층 모형
QUAD's 5th
floor model

The Taboo of Balance

When we come face-to-face with a challenge and overcome it, we often feel beauty within it. When we encounter something that is good but difficult to accomplish and so rare, we think of uniqueness, freshness, and rarity. While not all will agree, but for us, relative scarcity is closer to beauty than geometric rules like the golden ratio. The beauty we mean here is deeply connected to the civilization of this era, the economic structure of this society, and the depth of technological advancement.

When it comes to life, we don't stop imagining because we want to find the good things to create. The ongoing imagination gives us arduous tasks as life goes on, and we stubbornly tackle each one. That's how we get to the beauty we have in mind.

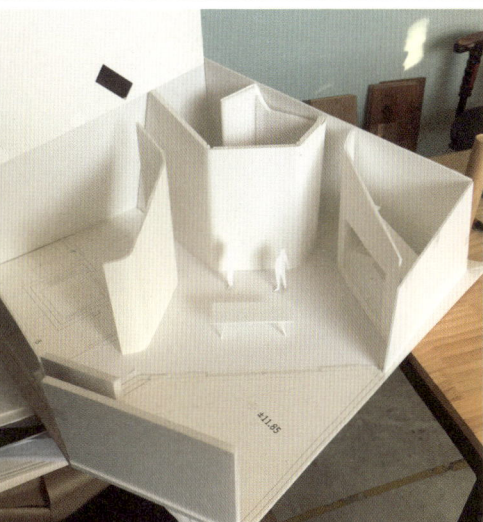

김남건축 | KimNam Architects

65 | The Taboo of Balance | 균형이라는

프로젝트

프라콩뒤 주택
- 위치: 프라콩뒤길, 오뜨-난다, 발레, 스위스
- 용도: 단독주택
- 설계담당: 김진휴, 남호진
- 대지면적: 1,994㎡
- 건축면적: 124.8㎡
- 연면적: 204.7㎡
- 건폐율: 6.26%
- 용적률: 10.27%
- 규모: 지하 1층, 지상 2층
- 구조: 철근콘크리트
- 외부마감: 목재, 노출콘크리트
- 설계기간: 2012. 02. ~ 2014. 08.
- 시공기간: 2014. 08. ~
- 구조설계: 슈넷처푸스카스 엔지니어
- 기계설계: 라파엘 푸르니에
- 시공: 건축주 직영
- 사진: 김남건축

르클루 레스토랑
- 위치: 클루길, 오뜨-난다,발레, 스위스
- 용도: 음식점, 게스트하우스
- 설계담당: 김진휴, 남호진
- 대지면적: 501㎡
- 건축면적: 68㎡
- 연면적: 174㎡
- 건폐율: 13.57%
- 용적률: 27.15%
- 규모: 지하 1층, 지상 2층
- 구조: 목구조, 철근콘크리트
- 외부마감: 목재, 노출콘크리트
- 설계기간: 2012. 02. ~
- 시공기간: 2013. 01. ~
- 시공: 건축주 직영
- 사진: 김남건축

쿼드
- 위치: 서울특별시 강남구 신사동
- 용도: 다세대주택, 근린생활시설
- 설계담당: 김진휴, 남호진
- 대지면적: 319.71㎡
- 건축면적: 181.03㎡
- 연면적: 559.79㎡
- 건폐율: 56.62%
- 용적률: 187.60%
- 규모: 지상 5층
- 구조: 철근콘크리트
- 외부마감: 칼라강판, 스터코, 포천석, 노출콘크리트
- 설계기간: 2017. 02. ~ 2018. 01.
- 시공기간: 2018. 03. ~ 2019. 04.
- 구조설계: 터구조
- 기계설계: ㈜기술사사무소 타임테크
- 전기설계: ㈜극동파워테크
- 시공: 엠오에이종합건설㈜
- 사진: 김남건축, 김경태

케이엔오
- 위치: 서울특별시 강남구 논현동
- 용도: 업무시설
- 설계담당: 김진휴, 남호진
- 대지면적: 276.6㎡
- 건축면적: 137.1㎡
- 연면적: 880.4㎡
- 건폐율: 49.57%
- 용적률: 247.83%
- 규모: 지하 1층, 지상 5층
- 구조: 철근콘크리트
- 외부마감: 포천석
- 설계기간: 2018. 09. ~ 2019. 01.
- 시공기간: 2019. 02. ~ 2020. 01.
- 구조설계: 터구조
- 기계설계: ㈜기술사사무소 타임테크
- 전기설계: ㈜극동파워테크
- 시공: (주)코워커스
- 사진: 김경태

이의동 상가주택
- 위치: 경기도 수원시 영통구 이의동
- 용도: 다가구주택, 근린생활시설
- 설계담당: 김진휴, 남호진
- 대지면적: 318.1㎡
- 건축면적: 183.72㎡
- 연면적: 539.32㎡
- 건폐율: 57.76%
- 용적률: 169.54%
- 규모: 지상 4층
- 구조: 철근콘크리트
- 설계기간: 2019. 10. ~ 2020. 11.
- 시공기간: 2021. 02. ~ 2022. 01.
- 구조설계: 윤구조기술사사무소
- 기계설계: ㈜기술사사무소 타임테크
- 전기설계: ㈜극동파워테크
- 시공: 엠오에이종합건설㈜
- 사진: Texture on Texture, 김남건축

Lighter Than Colors
- 위치: 광주광역시 북구 우산동
- 용도: 운동시설 (볼링장)
- 설계담당: 김진휴, 남호진, 이유나
- 연면적: 1,114.7㎡
- 설계기간: 2020. 04. ~ 2020. 07.
- 시공기간: 2020. 08. ~ 2020. 09.
- 시공: 무하
- 사진: 김남건축

이 도시의 유치원
- 위치: 서울특별시 동작구 흑석동
- 용도: 교육연구시설(유치원)
- 설계담당: 김진휴, 남호진, 유승주
- 대지면적: 21,550㎡
- 건축면적: 1,399.95㎡
- 연면적: 2,306.41㎡
- 건폐율: 6.49%
- 용적률: 9.89%
- 규모: 지하 1층, 지상 3층
- 구조: 철근콘크리트
- 외부마감: 벽돌
- 설계기간: 2020. 07.(현상) / 2020. 09. ~ 2021. 06.
- 시공기간: 2021. 09. ~ 2022. 12.
- 구조설계: 윤구조기술사사무소
- 기계설계: 타임테크
- 전기설계: ㈜극동파워테크
- 시공: ㈜에스앤비건설
- 사진: 김남건축

Mimosa
- 위치: 서울특별시 강남구 압구정동
- 용도: 근린생활시설(의원)
- 설계담당: 김진휴, 남호진, 이유나
- 연면적: 2㎡
- 설계기간: 2020. 07. ~ 2020. 07.
- 시공기간: 2020. 08. ~ 2020. 09.
- 사진: 김남건축

호숫가의 집
- 위치: 충청북도 제천시 청풍면 양평리
- 용도: 단독주택
- 설계담당: 김진휴, 남호진, 조경학
- 대지면적: 999㎡
- 건축면적: 198.66㎡
- 연면적: 238.9㎡
- 건폐율: 19.89%
- 규모: 지하 1층, 지상 1층
- 구조: 철근콘크리트
- 외부마감: 목재, 호피석
- 설계기간: 2021. 10. ~ 2022. 03.
- 시공기간: 2022. 05. ~ 2023. 06.
- 구조설계: 윤구조기술사사무소
- 기계설계: 서인엠이씨
- 전기설계: ㈜극동파워테크
- 시공: 무일건설㈜
- 사진: 김남건축

Projects

Chalet à Pracondu
- Location: Route de Pracondu, Haute-Nendaz, Valais, Switzerland
- Program: Single family house
- Architects: Jinhyu Kim, Hojin Nam
- Site area: 1,994m²
- Building area: 124.8m²
- Gross floor area: 204.7m²
- Building coverage ratio: 6.26%
- Floor area ratio: 10.27%
- Number of levels: B1, 2F
- Structure: Reinforced concrete
- Exterior Finish: Timber, Fairfaced Concrete
- Design Period: 2012. 02. – 2014. 08.
- Construction Period: 2014. 08. –
- Structural engineering: Schnetzer Puskas Ingenieure
- Mechanical engineering: Raphaël Fournier
- Construction: Owner-Builder
- Photography: KimNam Architects

Restaurant Le Clou
- Location: Chemin du Clous, Haute-Nendaz, Valais, Switzerland
- Program: Restaurant & Guesthouse
- Architects: Jinhyu Kim, Hojin Nam
- Site area: 501m²
- Building area: 68m²
- Gross floor area: 174m²
- Building coverage ratio: 13.57%
- Floor area ratio: 27.15%
- Number of Levels: B1, 2F
- Structure: Timber Frame Structure, Reinforced Concrete
- Exterior Finish: Timber, Fairfaced Concrete
- Design Period: 2012. 02. –
- Construction Period: 2013. 01. –
- Construction: Owner-Builder
- Photography: KimNam Architects

QUAD
- Location: Sinsa-dong, Gangnam-gu, Seoul, Republic of Korea
- Program: Multi-Unit Housing, Neighborhood Facility
- Architects: Jinhyu Kim, Hojin Nam
- Site area: 319.71m²
- Building area: 181.03m²
- Gross floor area: 559.79m²
- Building coverage ratio: 56.62%
- Floor area ratio: 187.60%
- Number of Levels: 5F
- Structure: Reinforced concrete
- Exterior Finish: Sheet Metal, Stucco, Stone Panel(Granite), Fairfaced Concrete
- Design Period: 2017. 02. – 2018. 01.
- Construction Period: 2018. 03. – 2019. 04.
- Structural engineering: Teo Structure
- Mechanical engineering: Timetech
- Electrical engineering: Keukdong Power Tech Co. Ltd
- Construction: M.O.A. Construction Co. Ltd
- Photography: KimNam Architects, Kyoungtae Kim

KNO
- Location: Nonhyun-dong, Gangnam-gu, Seoul, Republic of Korea
- Program: Office
- Architects: Jinhyu Kim, Hojin Nam
- Site area: 276.6m²
- Building area: 137.1m²
- Gross floor area: 880.4m²
- Building coverage ratio: 49.57%
- Floor area ratio: 247.83%
- Number of Levels: B1, 5F
- Structure: Reinforced concrete
- Exterior Finish: Stone Panel(Granite)
- Design Period: 2018. 09. – 2019. 01.
- Construction Period: 2019. 02. – 2020. 01.
- Structural engineering: Teo Structure
- Mechanical engineering: Timetech
- Electrical engineering: Keukdong Power Tech Co. Ltd
- Construction: Coworkers Co. Ltd
- Photography: Kyoungtae Kim

Curve Cut Sharply
- Location: Iui-dong, Yeoungtong-gu, Suwon, Gyeonggi-do, Republic of Korea
- Program: Multi-Unit Housing, Neighborhood Facility
- Architects: Jinhyu Kim, Hojin Nam
- Site area: 318.1m²
- Building area: 183.72m²
- Gross floor area: 539.32m²
- Building coverage ratio: 57.76%
- Floor area ratio: 169.54%
- Number of Levels: 4F
- Structure: Reinforced Concrete
- Exterior Finish: Brick
- Design Period: 2019. 10. – 2020. 11.
- Construction Period: 2021. 02. – 2022. 01.
- Structural engineering: Yoon Structural Engineers
- Mechanical engineering: Timetech
- Electrical engineering: Keukdong Power Tech Co. Ltd
- Construction: M.O.A. Construction Co. Ltd
- Photography: Texture on Texture, KimNam Architects

Lighter Than Colors
- Location: Usan-dong, Buk-gu, Gwangju, Republic of Korea
- Program: Sports Facility(Bowling Lane)
- Architects: Jinhyu Kim, Hojin Nam, Youna Lee
- Gross floor area: 1,114.7m²
- Design Period: 2020. 04. – 2020. 07.
- Construction Period: 2020. 08. – 2020. 09.
- Construction: Muha Co. Ltd
- Photography: KimNam Architects

The Kindergarten of the City
- Location: Seodal-ro, Dongjak-gu, Seoul, Republic of Korea
- Program: Educational Facility(Kindergarten)
- Architects: Jinhyu Kim, Hojin Nam, Seungju Yoo
- Site area: 21,550m²
- Building area: 1,399.95m²
- Gross floor area: 2,306.41m²
- Building coverage ratio: 6.49%
- Floor area ratio: 9.89%
- Number of Levels: B1, 3F
- Structure: Reinforced concrete
- Exterior Finish: Brick
- Design Period: 2020. 07.(competition) / 2020. 09. – 2021. 06.
- Construction Period: 2021. 09. – 2022. 12.
- Structural engineering: Yoon Structural Engineers
- Mechanical engineering: Timetech
- Electrical engineering: Keukdong Power Tech Co. Ltd
- Construction: S&B Construction Co. Ltd
- Photography: KimNam Architects

Mimosa
- Location: Apkujung-dong, Gangnam-gu, Seoul, Republic of Korea
- Program: Dermatology Clinic
- Architects: Jinhyu Kim, Hojin Nam, Youna Lee
- Gross floor area: 2m²
- Design Period: 2020. 07. – 2020. 07.
- Construction Period: 2020. 08. – 2020. 09.
- Photography: KimNam Architects

A House by a Lake
- Location: Yangpyeongri, Jecheon, Chungcheongbuk-do, Republic of Korea
- Program: Single-Family House
- Architects: Jinhyu Kim, Hojin Nam, Gyeonghack Jo
- Site area: 999m²
- Building area: 198.66m²
- Gross floor area: 238.9m²
- Building coverage ratio: 19.89%
- Number of Levels: B1, 1F
- Structure: Reinforced concrete
- Exterior Finish: Timbersiding, Stone Panel(Golden Yellow)
- Design Period: 2021. 10. – 2022. 03.
- Construction Period: 2022. 05. – 2023. 06.
- Structural engineering: Yoon Structural Engineers
- Mechanical engineering: Seoin MEC
- Electrical engineering: Keukdong Power Tech Co. Ltd
- Construction: Mooil Construction Co.Ltd
- Photography: KimNam Architects

리뷰 건축의 내재적 가치와 외재적 조건

남성택

젊은 건축가상은 '창의적으로 역량 있는 젊은 건축가들의 발굴'을 목표로 한다. 그러므로 영예로운 상이다. 다만 기성 건축가와 같은 완숙함을 지향하는 성인식이 되어서는 안 된다. 오히려 더 젊은 건축가가 되겠다는 다짐의 순간이어야 한다. 역사 속 위대한 건축가들은 끊임없이 자신을 쇄신했다. 영원히 젊은 건축가들이었다.

그러므로 김남건축의 건축가들에게 축하와 더불어 한마디 덧붙인다면, 계속해서 더 젊어지는 건축가가 되시라 바랄 뿐이다. 스스로 쇄신할 수 있는 방법 중 내가 알고 있는 것이 하나 있다. 그동안 살아온 건축의 여정을 원점의 순간까지 되짚으며 거슬러 올라가 보는 것이다. 건축가가 살아온 결정적인 순간들이 그가 발전시켜온 건축적 사고와 태도, 문제의식에 적지 않은 영향을 끼친다고 나는 믿는다. 그러므로 김남건축의 건축을 바로 분석하기보다 그 건축가들이 거쳐 온 삶의 궤적을 내 나름의 방식으로 조망하면서 글을 시작하려고 한다. 비록 제3자에 의한 일부 허구적 해석도 포함될 수도 있겠으나, 이러한 방식을 통해 그들의 핵심적 문제의식을 추론하거나 구축해 볼 수 있고 또 그들의 작품들에 대한 새로운 시각도 가능할 수 있을 것이라고 생각한다.

뉴헤이븐과 아이젠만

김진휴와 남호진이 처음 만난 곳은 미국의 뉴헤이븐이었다. 서울대와 이화여대에서 각각 건축을 공부한 후 미국으로 유학을 떠났고 그렇게 예일대 캠퍼스에서 마주치게 된 것이다. 예일대는 여느 미국 동부 건축 학교들처럼 파리 보자르 교육 체계를 수용하며 발전했다. 회화나 조각처럼 건축은 전위 예술로 인식되었고 이러한 관점이 예일대 건축의 DNA로 각인되었다. 2차 세계대전 전후 미국에서 근대건축의 영향이 커졌고 이에 대한 비판적 태도도 발전했음을 우리는 안다. 예일대는 그 대안적 전위 건축 운동들의 산실 중 하나였다. 브루탈리즘, 메가스트럭처와 연관된 루이스 칸(Louis Kahn)과 폴 루돌프(Paul Rudolph), 포스트모더니즘을 유행시킨 로버트 벤추리(Robert Venturi)와 찰스 무어(Charles Moore) 등 예일대를 거쳐간 교수들의 이름을 통해 이를 짐작할 수 있다.

당시 김남건축의 건축가들에게 깊은 인상을 남긴 선생은 단연 피터 아이젠만(Eisenman)이었다. 그들은 2001년부터 예일대에 합류한 그의 세미나 수업 'Introduction to Visual Studies'에서 건축 사례들의 형태적 체계를 발견하고 분석한 핵심을 명확한 다이어그램으로 시각화하는 법을 익혔다.그림1 아이젠만은 자타공인 형태주의자였다. 그의 1963년 박사 논문은 근대건축 사례들의 '형태 체계' 연구였고 이 이론은 1968년부터 시작된 주택 시리즈 설계 연구로 이어졌다. 특히 그는

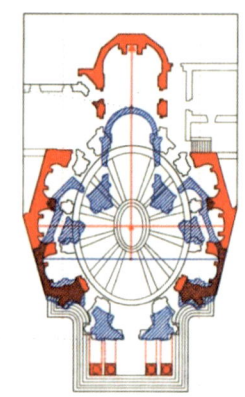

그림 1 피터 아이젠만 세미나 수업 'Introduction to Visual Studies' 과제

Review
Architecture's Intrinsic Values and Extrinsic Conditions
Sungtaeg Nam

The Korean Young Architect Award aims at "hunting creatively capable young architects," so that it is obviously an honor for its recipients. However, it should not be like a coming-of-age ceremony to seek after some maturity of established architects, but rather be the moment of enabling resolutions to be fresher as architects. The great architects in history have ceaselessly endeavored to innovate themselves, so as to be recognized as permanently young architects.

Hence, the one thing I would say with celebration to the architectural duo of KimNam Architects is that I hope you will keep getting younger as architects. As far as I know, one of many possible ways to innovate yourselves is backtracking to the original moment of your architectural journeys you have walked through to date. I believe the decisive moments of an architect considerably influence his or her development of architectural thinking, attitude, and critical mind. In this sense, I will not enter a point-blank analysis of the duo's works but begin in my own speculation with the trajectories of their own lives. Although it could include some fictional interpretation by a third person, this approach will make it possible to infer or reconstruct their core critical minds as well as to understand their works from a new perspective.

New Haven and Eisenman
Jinhyu Kim and Hojin Nam first met each other in New Haven, Connecticut, USA. Upon graduating in architecture from Seoul National University and Ewha Womans University respectively, they went to study abroad in the United States and encountered each other at the Yale campus. Like any architecture schools in the eastern US, the Yale School of Architecture has also developed by accommodating the education system of École des Beaux-Arts. Architecture was recognized as a vanguard art like painting or sculpture, and this perspective was inscribed as the Yale architecture school's DNA. We are well-informed that the influence of modern architecture increased in the United States around the 2nd World War along with the growth of criticisms against it. At the time, Yale was one of the cradles of such alternative vanguard architectural movements, as evidenced by the names of its former professors like Louis Khan and Paul Rudolph, who were associated with brutalism and megastructure, and Robert Venturi and Charles Moore, who made postmodernism into a fashion.

One of the most impressive teachers to Kim and Nam at the time was obviously Peter Eisenman who, since he joined Yale in 2001, opened a seminar named "Introduction to Visual Studies" in

1980년대 초 해체주의(Deconstructivism) 건축 운동에 참여했고, 1990년대 초 전자적 패러다임 아래 디지털 건축의 형태적 가능성을 탐구했다. 이 같은 다양한 활동에도 아이젠만의 건축은 본질적으로 형태였다. 그 외의 조건, 요소들은 의도적으로 배제하거나 하부 문제로 간주했다.

김남건축이 아이젠만으로부터 계승한 것이 있다면 단지 순수한 형태주의만은 아니었다. 형태를 포함해 건축 고유의 자율적 가치와 내재성이 중요하다는 본질주의적 태도였다. 김남건축에게 형태 논리도 물론 중요했지만 구조, 구축, 공간 등 다른 건축적 가치도 중요했다. 형태만 건축의 내재적 가치인 것은 아니었다.

바젤과 헤르조그 & 드뫼롱

김남건축의 확장적 태도는 김진휴, 남호진이 스위스로 옮기면서 더욱 분명해진다. 헤르조그 & 드뫼롱 밑에서 실무할 기회가 주어지자 대서양 너머 바젤로 길을 떠나기로 한 것이다. 새로운 생활 환경에 적응하는 것도 힘들었겠지만, 건축적 풍토도 미국과는 매우 달랐다.

뉴헤이븐이 형태 중심 전위 건축의 중심지 중 하나였다면, 바젤은 '구축(Bau)'을 중시하는 스위스 건축의 메카였다. 바젤은 뛰어난 건축가들을 계속 배출하고 있을 뿐만 아니라, 기념비적 건축들도 즐비했다. 더 나아가 일상의 건축, 심지어 변압소나 창고 같은 산업용 건물조차 건축작품이 될 수 있는 건축 도시였다.

바젤은 예술 도시이기도 했다. 단지 '아트 바젤'의 도시여서만이 아니라 좋은 미술관들과 예술 컬렉션들이 갖춰져 있고, 창의적인 예술가들의 활동 무대이기도 했다. 자크 헤르조그(Jacques Herzog)와 피에르 드뫼롱(Pierre de Meuron)은 요제프 보이스(Joseph Beuys)의 그 유명한 1978년 바젤 카니발을 보조하는 작업으로 데뷔했으며, 독일 예술가로부터 재료의 내재적 물성을 배웠다. 같은 해 헤르조그 & 드뫼롱 설계사무실을 개소하며 선보였던 그들의 건축은 미니멀 아트나 팝아트처럼 형태적 구성예술에 반대한 비구성(non-composition) 예술의 문제의식에 기반한 작업이었다. 그들의 소위 상자 건축은 미국 포스트모던 및 해체주의 건축에 대한 직접적 항쟁이기도 했다. 근대건축의 획일적 형태를 비판하며 지역과 맥락, 상징적 형태의 부활(포스트모던)을 주장하거나 파편, 추상, 탈구축적 형태의 재창조(해체주의)를 주장하는 것은 본질적으로 여전히 형태주의들 간의 논쟁일 뿐이라는 것이다. 최소한 1990년대 초까지 이들은 가장 무관심한 형태로 한정 지으며 형태주의적 해석의 개입을 원천 차단했고, 구축과 인지현상의 문제에만 집중하고자 했다.

헤르조그 & 드뫼롱은 아이젠만의 대척점과 같았다. 이러한 점에서 김남건축의 유학과 실무는 상반된 이데올로기 간의 사상전환처럼 보일 수도 있겠다. 개념적, 형태적 미국 건축에서 현상학적, 구축적 스위스 건축으로의 전향처럼 말이다. 물론 김남건축의 경험과 연관된 21세기 초에는 아이젠만이 현상학적 건축(베를린 홀로코스트 메모리얼)을 실현했고, 헤르조그 & 드뫼롱은 독창적인 형태(베이징올림픽 스타디움)를 선보였던

which they learned how to visualize the core discovery and analysis of the formal system of architectural cases into clear diagrams.[Figure 1] Eisenman was recognized as a formalist by everyone including himself. His doctoral dissertation in 1963 was a study on the "formal system" of the cases of modern architecture, and this theory triggered his research of house design series since 1968. Particularly, he joined in the deconstructivist movement of architecture in the early 1980, and explored the formal possibilities of digital architecture within the electronic paradigm of the early 1990s. For all these various activities, his architecture was essentially based on "form." Other conditions and elements were intentionally excluded or considered subordinate matters.

What KimNam inherited from Eisenman, if any, was not the pure formalism, but the essentialist attitude of valuing autonomous values and immanence proper to architecture including form. The formal logic was important to KimNam as a matter of course, but other architectural values like structure, construction, and space also mattered to the duo. The form was not the only value immanent to architecture.

Basel and Herzog & de Meuron
KimNam's comprehensive attitude would become clearer as they moved to Switzerland. With the chance given to work at the practice of Herzog & de Meuron, they decided to be on the way to Basel over the Atlantic Ocean. Not only adapting to the new living environment would have been challenging, but the architectural culture itself was very different from that of the United States.

New Haven was one of the centers of form-based vanguard architecture, while Basel was a mecca of Swiss architecture valuing "construction" (Bau). Basel was not only producing excellent architects continuously, but also lined with rows of monumental buildings like museums. Moreover, it was the city where everyday buildings, even industrial buildings like power substations and warehouses could become architectural works.

Basel was also an artistic city, not only because it held "Art Basel" but also because it was well-equipped with good museums and collections, serving as an activity platform for creative artists. Jacques Herzog and Pierre de Meuron made a debut by assisting the famous 1978 Basel Carnival works of Joseph Beuys, the German artist from whom they learned about immanent materiality. Their early works shown with the inauguration of their practice in the same year were based on a non-compositionist critical mind, as shown in minimal or pop arts, against formal compositionist arts. Herzog & de Meuron's so-called boxy architecture was also a direct strife against the American postmodern and deconstructivist architectures. Which was to say that asserting the (postmodern) resuscitation of localities, contexts, or symbolic forms or the (deconstructivist) recreation of fragmentary, abstract, or deconstructed forms in the name of criticizing the modern architecture essentially remains a formalist polemic. Until the early 1990s at the latest, the Swiss architects foreclosed the formal interpretation by limiting their use of forms to the most indifferent ones, to focus on such matters as constructions and cognitive phenomena.

시기였다. 영원히 동떨어져 있을 것 같던 두 행성이 순간적으로 교차 정렬한 것처럼 보일 수도 있는 때였다. 물론 애초부터 두 건축은, 시각적 형태의 건축이든 공감각적 현상학과 구축의 건축이든, 모두 건축의 자율성과 내재성을 중시해왔다는 내면적 공통점도 공유하고 있었다. 바로 이 접점 위에 김남건축이 이어지고 있는 것일까.

알프스와 샬레, 건축가 없는 건축의 건축가

2012년 김진휴와 남호진은 바젤에서의 커리어를 중단하고 스위스 불어권 알프스의 산골마을에 자리잡는다. 유명 스키장 지역인 오트-난다의 프라콩뒤 마을에 작은 오두막집을 설계해달라는 제안을 받은 것이다. 의뢰인은 바젤에서 알게 된 목수로, 그는 오래된 샬레(산장)들이 사라지는 것이 안타까워 사 모으기 시작했고, 샬레 중 일부를 이용해 자기 집으로 개조하고 싶어했다. 두 건축가는 건축부지 옆에 살며 작업하기로 마음먹는다. 이때 김남건축이 설립된다. 알프스 산골마을에서 시작된 것이다.

　　김남건축은 현학적 건축의 세계를 떠나 토속적 건축인 샬레를 마주하게 된다.^{그림2} 샬레는 유럽 산악 목조건축을 대표하는 민중건축의 한 유형이다. 최소 외피로 최대 공간을 구축할 수 있도록 단순하고 조밀한 입방체 볼륨을 지향한다. 구축적, 공간적, 환경적 측면에서도 효율적인 형태이다. 내부는 다양한 기능을 포용한다. 산골의 주거나 피난처로 쓰이고, 치즈의 생산 및 숙성 장소로도 사용되며, 가축을 키우거나 건초를 보관하는 창고로 쓰이기도 한다. 중립적 공간의 구조체인 것이다. 샬레의 형태는 구축적 논리에 따른 결과이다. 물론 주요 부분은 주변에서 쉽게 구할 수 있는 목조로 이루어진다. 하지만 하부는 겨울철에 눈으로 파묻히는 부분이어서 상당한 높이까지 괴석을 쌓아 구축되는 토대 위에 세워진다. 결국 엄밀하게 말하면 샬레는 석재 하부와 목재 상부가 결합된 복합 구조체다. 최상단은 샬레의 상징적인 형태인 큰 맞배지붕으로 마무리된다. 지역의 기후에 대응해 지붕 경사가 급하고 처마는 길게 뻗어 있다. 지붕면은 나무 판재로 마감되거나 오랫동안 눈으로 덮여 있어야 할 때를 대비해 자연 판석들로 보호된다.

　　샬레의 형태는 각 부분이 개별적 필요에 독립적으로 대응한 것으로, 특별한 원칙 없이 전체가 결합된 효율적 구축의 결과이다. 산골사람들이 저마다의 필요와 수단에 따라 구축하며 발전시킨 건축물인데도 하나의 공통적 유형(type)으로 귀결될 정도로 샬레의 형태는 표준화되었다. 동일한 화학적, 물리적 조건에서 생성되는 수정체들의 형태처럼 말이다. 동시에 샬레는, 원래 유형의 정의가 그런 것처럼 비슷하되 획일적이지는 않다. 샬레의 체계는 비교적 명확하나 얼마든지 형태적, 구축적 융통성을 지닐 수 있다. 그럼에도 불구하고 이웃한 샬레들은 서로 잘 어울린다. 주변 자연과의 위화감도 없다. 원래 그 땅에서 태어난 생명체처럼 그냥 자연스

그림 2 스위스 샬레

Herzog & de Meuron was like an antithesis to Eisenman. In this sense, Kim and Nam's works in school and practice might look like an intellectual conversion between opposite ideologies, as if they converted from the conceptual and formal American architecture to the phenomenological and constructional Swiss architecture. Still, the early 21st century related to Kim and Nam's training experiences saw Eisenman realizing phenomenological architecture (*i.e. the Holocaust Memorial in Berlin*) and Herzog & de Meuron displaying an original form (*i.e. the Olympic Stadium in Beijing*), as if the two planets which had seemed like being parted forever momentarily became trans-arranged. It should also be noted that both architectures have valued a common feature inside from the beginning: the autonomy and immanence of architecture, be they visually formal or synesthetically phenomenological and constructional. At this very intersection might KimNam stand inheriting them?

The Alps and Chalets, and the Architect of "Architecture Without Architects"
In 2012, Kim and Nam stopped their careers in Basel and settled in a Francophone Alpine village in Switzerland, when they were commissioned to design a small hut in the village of Pracondu, Haute-Nendaz, one of the famous ski areas. The client was a carpenter whom the duo had known in Basel. Feeling pity for the old chalets disappearing, he had begun to buy and collect them and decided to renovate some of them into his own houses. For this design task, the architect duo decided to reside by the site, even as they started up their independent practice in the Alpine village.

Leaving the pedantic architectural world, KimNam directly faced the vernacular architecture of chalets.^{Figure 2} A chalet is a type of folk architecture representing the European mountainous wooden architecture. It pursues a cubic volume which is so simple and compact as to make it possible to construct maximum space with minimum envelope. It is also an efficient form in terms of construction, space, and environment. Its interior accommodates various functions, such as a house or a shelter in the mountainous village; a place for producing and ripening cheese; and a warehouse to raise livestock or store hay. A chalet is a construction of neutral space, and its form is the result of a constructional logic. Although its main part is made up of wood which is easily available from nearby areas, the lower part is built on the base constructed by stacking oddly-shaped stones to a considerable height. Strictly speaking, a chalet has a complex structure in which the lower stone and the upper wood are combined. The uppermost part is finished with a gable roof, which is the symbolic form of the chalet. The roof has a steep pitch and long eaves to cope with the local climate, and its surface is finished either with wooden boards or natural flagstones in case it is covered with snow for a long time.

럽다. 더 나아가 아름답다. 르 코르뷔지에는 알프스 산악건축에 원시건축과 같은 고귀한 기하학적 질서가 있다고 선언했다. 아돌프 로스는 시골 건축들이 원초적 본능에 따라 지어졌을 뿐인데도 선조들의 옛 건축과 조화를 이루며, 또 자연과 합일되는 보편적 아름다움에 도달한다고 예찬했다.

김남건축도 마치 고대 폐허를 발굴하는 고고학자들처럼 <프라콩뒤 주택> 설계안의 전제조건인 샬레를 실측, 분석해야 했다. 의도된 바 없는 재료의 물성이나 뜻밖의 공간 효과에도 심취했다. 동양의 건축가들은 선입견 없이 순수한 눈으로 관찰했다. 바닷가에서 주운 정체 모를 무언가처럼, 샬레는 그 자체로 '발견된 오브제(found object)'였다. 건축가들은 "조용한 대화"의 순간이라 묘사했다. 낯선 곳에서 낯선 사물과의 대화는 자신과의 대화이다. 동굴이나 광야처럼 영도(zero degre)의 상태에서 비로소 스스로를 발견할 수 있다. 그들은 제도판이나 모니터 위에 그릴 수 있는 머릿속 이상을 투영하는 것이 아니라 발 딛고 서있어야 할 현실적 조건들에 집중했다. '건축가 없는 건축'의 건축가처럼 당면 문제들을 하나씩 해결하고자 했으며 철물 카탈로그와 같은 파편적 현실에 기반한 '브리콜뢰르(bricoleur)'로 변신했다.

김남건축은 목재 틈 사이로 빛이 스며드는 샬레 내부에 콘크리트 주거 박스를 삽입하고자 했다.^{그림3} 내부와 외부의 각 재료적 순수성 사이에 긴장감이 생성될 것이다. 사실 형태적으로나 구축적으로 둘은 긴밀해야 했다. 콘크리트는 태생적으로 나무 거푸집 속의 허공(보이드)이 응고된 고체(솔리드) 아니던가. 콘크리트 매스는, 마치 레이첼 화이트리드(Rachel Whiteread)의 설치 작품 <Ghost House>(1990)처럼 샬레의 빈 공간을 틀로 삼아 주조한 영혼적 물체와 흡사할 것이다. 콘크리트의 형태는 샬레라는 알껍질을 깨고 태어난 생명체가 될 것이다. 김남건축의 건축도 그 속에서 함께 태어났다.^{그림4}

그림 3 프라콩뒤 주택, 2012-, 목조 샬레 속 콘크리트 주거 볼륨 다이어그램

서울의 재발견, 그리고 소필지 다세대 주거

2014년 착공된 <프라콩뒤 주택>의 공사는 매우 더디었다. 십여 년이 지난 지금도 여전히 완공되지 못했다. 공사할 수 있는 기간이 짧고 부재들을 경제적으로 운송하기도 어려운 알프스 산악 건축의 특성 때문이었다. 물론 '알프스 건축(Alpine Architeuture)'의 시간 개념은 도시 건축의 것과 다른 것이리라. 알프스 자체가 오랜 세월을 거쳐

그림 4 레이첼 화이트리드(Rachel Whiteread), Ghost House, 1990

The form of a chalet follows how each part independently fulfills its individual needs, as a result of efficient construction in which all parts are combined with no special principle. The form of buildings developed by the villagers' constructions to their individual needs and means as it is, it has been standardized to the extent of resulting in a common type, just like the form of crystals generated in the same chemical and physical conditions. Even as chalets have a similar form, however, they are not uniformized, as this is the original definition of a type. The system of a chalet is comparatively clear, but can have formal or constructional flexibilities. Nevertheless, adjacent chalets mingle well with each other. There is no sense of disharmony with nature as well. They look natural as if they were creatures born of the land. Furthermore, they are beautiful. Le Corbusier declared the Alpine mountainous architecture has a noble geometric order, similar to primitive architecture. Adolf Loos praised rural architecture for making harmony with the old architecture of ancestors and reaches the universal beauty that unites with nature although they were built by primitive instinct.

Like the archeologists excavating ancient ruins, KimNam also had to survey and analyze the chalet as a precondition to design a *house in Pracondu*. They were also absorbed in unexpected materiality or spatial effects. The Eastern architects observed with pure eyes without prejudices. Like some unidentifiable object picked up on a beach, a chalet is itself a "found object." The architects described this as a moment of "quiet conversation." Having a conversation with a strange object at a strange place means one is conversing with oneself. It is in a zero-degree condition like a cave or wilderness that one can find oneself. The duo did not project their mental ideals to be drawn on a drafting board or a monitor, but focused on the realistic conditions where they should stand up on their feet. Like the architect of "architecture without architects," they intended to resolve the matters at hand and turned into bricoleurs based on the fragmentary realities similar to a hardware catalogue.

KimNam intended to insert a concrete housing box in a chalet where light permeates through the gaps between wooden planks.[Figure 3] Between the material purities inside and outside would occur a sense of tension. In fact, both had to be close-knit in terms of form and construction. Is not concrete inherently a solidification of the void in the wooden form? The concrete mass would be similar to the spiritual object cast with the void of the chalet taken as a frame, as in Rachel Whiteread's installation, <*Ghost House*>(1990). The concrete form would become a creature hatched out of the eggshell called "chalet." KimNam's architecture was also born of it together.[Figure 4]

Seoul Rediscovered, and Multi-family Housing on Small Lots

The *Chalet à Pracondu* started to be built in

매우 천천히 생성, 경화, 변형되는 지질학적 구축물이라면, 그 땅과 하나가 될 건축 역시 영속적으로 진행될 프로젝트여야 하지 않을까. 영원히 완성되지 않을 미완성의 설계는 오히려 순수한 건축 개념으로 작용할 수도 있고 이상적 다이어그램과 같은 강력한 잔상을 남긴다.

프라콩뒤의 샬레는 김남건축의 시작이었다. 이는 공사가 지연되며 서서히 망각되어 갈 수도 있겠지만, 반대로 계속 머릿속을 맴도는 유령처럼 하나의 건축적 영감으로 확립되며 지속적인 영향을 끼쳤을 수도 있다. 예를 들어 외부 자연과 내부 공간의 전이 관계에 집중했던 <프라콩뒤 주택>설계 전략은 귀국 후 김남건축이 설계한 국내 단독주택 사례들에서 은연중 나타난다. 제천 <호숫가의 집>이나 독신자를 위한 <P주택> 설계안에서 보이는 내외부 사이의 공간적 액자가 그렇다.

물론 단독주택은 현대의 한국 주거건축에서 대표적 유형이 되지 못하고 있다. 반면 김남건축의 작업에 새로운 활력을 불어넣었던 <Quad>와 같은 도심 소필지 다세대주거는 점점 확산되어 가는 양상이다. 서울 도심 양옥지역이나 외곽 택지개발 주거지역에서 용적률에 따라 4~5층 높이로 지어지곤 하는 소필지 다세대/다가구주거는 대단지 고층 아파트 유형과 함께 한국 주거유형학의 대표적 요소로 자리잡았다. 대단지 아파트 시장을 대형 건설사 및 설계사가 장악하고 있다면, 소필지 다세대주거는 집장사들이 주도하고 있지만 비교적 소규모 자본의 건축이므로 아틀리에형 건축가들에게도 어느 정도 기회가 열려 있다.

한국의 다세대주거 건물이 익명의 건축이란 점에서 샬레와 비교될 수도 있다. 물론 다세대주거는 자본주의적 욕망에 따라 수익을 중시하는 재테크 수단으로 전락해 저급한 소모품 같은 건물들로 양산되고 있어 건축의 퇴행적 현상처럼 인식되곤 한다. 그럼에도 현대 한국사회의 보편적 풍경을 형성하는 주거 유형인 것은 부정할 수 없다. 어쩌면 우리 사회의 민낯이 고스란히 드러나는, 우리 시대의 '건축가 없는 건축'인 셈이다. 불편한 현실이라도 눈을 돌려선 안된다. 다세대주거는 젊은 건축가들의 생존을 연명하는 세속적 일거리가 되고 있지만, '지금 여기', 우리만의 건축이 싹틀 수 있는 미지의 텃밭일 수도 있다. 어떤 제약적 조건 속에서도 건축적 의미는 도출될 수 있어야 한다. 농부가 밭을 탓하랴. '건축가 없는 건축'은 잘못이 없다. 부끄러움은 '건축 없는 건축가'에게 있다.

김남건축에게 <Quad>의 설계는 프라콩뒤에서처럼 또다른 현실의 자각이었다. <Quad>의 형태는 의도적으로 이질적이며 하나로 조율되고 통합되지 않는다. 각 부분들은 각자의 문제를 고민할 뿐 서로의 관계는 단절적이다. 각 부분은 독자적 해법들로서 단지 수직 중첩될 뿐이다. 서로 무관한 요소들이 우연히 병치된 '카다브르 엑스키(cadavre exquis)'가 소환된다.[그림5] 그럼에도 불

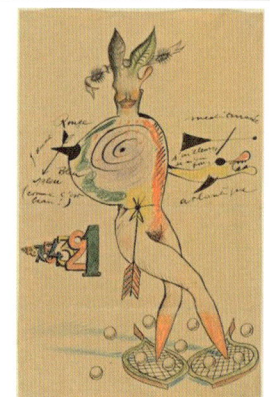

그림 5 카다브르 엑스키 (by Yves Tanguy, Joan Miro, Max Morise and Man Ray, 1926)

2014, but its works progressed too slowly. Almost a decade has passed since, but it was not completed to this day. This was because Alpine mountainous architecture put a challenging condition where the possible working period within a year was not long enough and the building components could not be shipped economically. Admittedly, the time scale of Alpine architecture will be different from that of urban architecture. If the Alps themselves are the geological constructions generated, solidified, and transformed very slowly for a long time, should not the architecture to become one with the site also be a project which will progress permanently? The incomplete design which will not be completed forever might rather serve as a pure architectural concept, leaving a strong afterimage like an ideal diagram.

The *Chalet à Pracondu* was the beginning of KimNam Architects. This origin might be forgotten slowly by the delay of construction, but also could have had a continuous influence as an architectural inspiration established like a ghost lingering in their heads. For example, the design strategy for the *Chalet à Pracondu*, which focused on the transition between the nature outside and the space inside, is implicitly reflected in the domestic detached houses that KimNam designed after coming back to Korea, such as the spatial frames between the inside and out as shown in *a House by a Lake* built in Jecheon or in the *P House* designed for bachelors.

A detached house is not a representative type in the present-day Korean residential architecture. However, the multi-family housing on a small urban lot, like the project *Quad* that brought new inspirations for KimNam's works, is increasingly wide-spreading. This type of multi-family or multi-unit housing on a small lot, often built as 4 to 5 stories by the floor area ratio in the western-style residential zone of downtown Seoul or the residential development of its outskirts, has settled as a representative element of the Korean residential typology along with the large-scale development of high-rise apartments. If the latter's market is dominated by large-scale contractors and architectural firms, the former's market is primarily led by commercial house-sellers but somewhat open to atelier-type architects as it is driven by comparatively small-scale capitals.

In terms of anonymous architecture, the Korean multi-family housing could be compared to the chalet. The multi-family housing is often recognized as a regressive phenomenon because it is mass-produced as a means of profit-based financial investment techniques following capitalist desires. However, this residential type is undeniably forming the universal landscape of the present-day Korean society, which is perhaps the "architecture without architects" of our time that reveals the bare face of our society just as it is. Although it is an uncomfortable reality, we should not turn our eyes. The multi-family housing is providing

구하고 일반 주거지역 소필지에 내재된 제약적 체계, 관습적 보편성에 기반하는 형태임은 부정할 수 없다. 김남건축은 이를 다세대주거 속에 각인된 "일종의 유전자"로 표현한다.

<Quad>의 분석을 통해 이를 한번 재정리해보자. 지상층은 주차를 위해 필로티 구조가 적용되며 상부층으로 올라가는 코어가 있고 작은 근린생활 사무공간을 포함할 수 있다. 주차와 근생은 도시적 프로그램이다. 사무공간의 내부 바닥은 충분한 천장고의 확보를 위해 낮춰질 수도 있다. 지상 9미터 이상 높이부터 적용되는 일조사선제한 때문에 각층 층고가 매우 제한되어 버린 부작용에 대한 절충적 해법이다. 따라서 지상층은 건축 가능한 매스로부터 속을 파내어 만드는 선사시대의 토굴 건축과 흡사해진다.

9미터 높이 이내인 2~3층은 부지의 건축선 한계 내에서 최대한의 사적 영토확장을 시도하는 자본주의적 영역이다. 주변에 대한 고려보다 내부 욕망으로 팽창하는 폐쇄적 블록이다. <Quad>에서 내부는 다양한 임대용 주거 평면이 퍼즐처럼 조합되어 있다. 결과적으로 창문들도 비규칙적으로 배치될 수 있다. 주변 건물의 개구부 위치까지 고려해 구성하는 섬세함도 발견된다. 다만 불안정한 도심에서 이와 같은 맥락적 디테일은 효용성의 수명이 한정적일 수 있다. 다세대주거에서 일반화될 수 있는 2~3층 부분의 특징으로는 '발견한 그대로의(as found)' 부지 형상에 따라 자동 압출(extrusion)된 프리즘으로 정리된다.

4층은 9미터 높이를 초과한 첫 번째 바닥층이다. 아래에서 단순했던 수직 입면선이 정북 방향에서 일조사선 적용으로 급격히 후퇴하게 된다. 결과적으로 비대칭적이거나 사선적인 윤곽이 생성되며 형태의 단순함은 위기를 맞이한다. 새로운 건축이 시작되는 플랫폼일 수도 있다. 한편 평평한 옥상 바닥이 강제된다는 사실에 주목할 필요가 있다. 물론 경우에 따라서 쓸모 없는 애매한 폭의 옥상 바닥도 생성될 수 있다. 테라스가 될 수 있지만 정북 방향이라는 한계가 있다. 그럼에도 불구하고 형태적 일탈과 외부 공간을 도입할 기회로 활용할 수 있다. <Quad>의 경우 지그재그의 픽처레스크 입면이 적용되며 테라스로 둘러싸인다. 결국 요약하면 4층은 테라스 하우스이다.

<Quad>에서 가장 시선을 끄는 부분은 5층이다. 마치 기단 위에 올려진 추상적 조각 같은 형상이 멀리서부터 부각된다. 펜트하우스가 마련되며 예상치 못한 내부 공간이 펼쳐진다. 다세대주거에서 최상층은 예외적인 부분이 될 수 있다. 하늘에 떠 있는 건축, '천공의 성'과 같은 초현실적 상상도 연출 가능하다. 최상층은 일조사선에 의해 윤곽이 완결되는 대리석 원석과 같은 것이다. 건축가는 조각가가 될 수 있다. 미켈란젤로의 표현처럼 "돌덩어리 안에 조각상이 숨겨져 있으며 그것을 발견하는 것이 조각가의 임무"인 것이다. 그림6

그림 6 Quad, 2017-2019

earthbound jobs for young architects to subsist on, but it could be an unknown field where our own architecture will spring up. From whatever restrictive conditions, we should be able to derive architectural meanings. Could a farmer blame the field? There is no fault in the "architecture without architects." The shame is on the "architects without architecture."

For KimNam, designing *Quad* was an experience of realizing another reality as was in Pracondu. The form of *Quad* is intentionally heterogeneous, not coordinated and integrated as one. Each part considers their own matters, with their interrelationship broken. Each part superposes with another part just vertically as an independent solution. Summoned here is the "cadavre exquis" where elements are juxtaposed accidentally with no interrelation.^{Figure 5} Nevertheless, it is undeniably the form based on conventional universality, the restrictive system inherent in a small lot of a general residential zone. KimNam expresses this as "a kind of genes" inscribed in the multi-family housing.

Let's take this again through the analysis of *Quad*. The ground floor has a piloti structure for parking, the core leading upstairs, and a small neighborhood office. The parking and the neighborhood office are urban programs. The internal floor of the office space can be lowered to gain a sufficient ceiling height. This is an eclectic solution to the side effect of considerably restricted ceiling heights due to the setback regulation which applies above 9 meters from the ground level. Thus, the ground floor becomes similar to an underground tunnel of the prehistoric ages that is made by hollowing out the inside of the buildable mass.

The 2nd to 3rd floors located within the height of 9 meters are the capitalist realm that attempts to expand private territories to the maximum within the limit of the building line. This realm is a closed block that expands with internal desires rather than with considerations about the surroundings. The interior of *Quad* consists of various floor plans of rental living units combined like a puzzle. In result, the windows can be arranged irregularly. The subtlety of considering where the openings of adjacent buildings are located is also found. However, these contextual details could be short-lived in the unstable urban center. The 2nd to 3rd floors, which can be generalized as a type for multi-family housing, are characterized by a prism extruded automatically according to the "as-found" figure of the site.

The 4th floor is the first floor above 9 meters from the ground. The façade line which is simply vertical at the lower part becomes radically withdrawn from the due northern direction according to the setback regulation. In result, an asymmetrical or oblique profile occurs and the simplicity of form faces a crisis. It could be a platform where a new architecture begins. On the other hand, it should be noted that a flat roof space is introduced forcibly. In some cases, this could create a useless and ambiguous

해부대 위의 재봉틀과 우산의 우연한 만남

<Quad> 이후 김남건축은 여러 다세대주거 시리즈를 동시 설계하게 된다. 기본적인 체계의 유형은 공유하되 다채로움을 부여할 수 있는 개별적 시도가 이루어진다. 구조적 다양성의 실험까지 추가된다. 최근 완공된 일원동 다세대주거 <Warm and Cool>은 일상의 건축이 도달할 수 있는 전위성의 한계에 도전한다.^{그림7}

지상층에는 외부 기둥을 없애고자 상당한 경간의 캔틸레버 구조를 과감하게 적용했다. 수직 구조체는 한 가운데 좁게 세워진 철근콘크리트 내력벽 볼륨뿐으로 검정색으로 채색되어 존재감을 숨긴다. 내부에는 근생 사무공간이 포함되어 있는데, 역시 내부 바닥을 낮추어 천장고를 확보했다. 캔틸레버 덕분에 사무공간 깊숙이 투명한 모퉁이 창이 만들어질 수 있어 햇살이 내부로 쏟아진다.

2~3층은 최대한 넓게 펼쳐진 면적으로 이루어진 캔틸레버 볼륨이다. 내부 평면은 원룸주거가 테트리스처럼 결합된다. 커다란 격자창이 규칙적으로 뚫린 입면은 철근콘크리트 비렌딜 트러스(vierendeel truss) 그 자체이다. 혹시 모를 바닥처짐을 방지하기 위한 결정이었다. 사선이 아닌 직각, 힌지접합이 아닌 강접합 구조인 비렌딜 트러스는 구조역학적인 단점이 없지 않지만, 공간적 효용성을 중시하는 건축에서 특별한 쓸모를 발견한다. 물론 소규모의 일상 건축에서 적용된 사례는 드물었는데 여기서는 지상층 주차의 편리함을 위해 정당화된다.

<Warm and Cool>의 건축 부지는 넓은 면이 정북 방향이어서 일조사선의 영향이 컸다. 따라서 4층 이상에서는 남쪽에 치우친 좁고 긴 볼륨만 구축 가능했다. 4층과 5층은 4인 가족을 위한 펜트하우스이다. 뜻밖에도 여기서는 철골 구조가 쓰였는데, 좁은 볼륨에서 공간을 절약할 수 있다는 명분으로 적용되었다. 실제로는 석조와 목조가 결합된 샬레처럼, 여기서도 구조적 복합성을 실험하려는 건축가의 숨겨진 야망과 즐거움도 엿보인다. 4층은 북쪽 테라스와 도로 쪽 정면을 향한 전면창이 둘러싸고 거실과 식당이 통합된 하나의 열린 공간으로 되어 있다. 반면 5층은 침실들로 구성되어 있고 폐쇄적인 입면을 지니며 단순 맞배지붕으로 마무리된 닫힌 공간으로 요약될 수 있다. 가족 공동체를 위한 낮의 공간과 개개인을 위한 밤의 공간으로 구분되는데, 필립 존슨(Philip Johnson)의 유리주택이 보여주는 극단적 투명성이 그 옆 벽돌주택의 완강한 불투명성 덕분에 보완될 수 있었던 것처럼, 대조적이면서 상호보완적인 공존이 여기서 발견된다.

결과적으로 일원동 다세대주거의 생경한 형태는 로트레아몽(Lautréamont)의 시구이자 초현실주의 공식이 된 문구 "해부대 위의 재봉틀과 우산의 우연한 만남"과 같다. 외관상 건축의 형태는 미스테리다. 좁다란 블랙박스 위에 올려진 비렌딜 트러스가 시소 마냥 길게 양쪽으로 펼쳐진다.

그림 7 일원동 다세대주거, 2019~2022
ⓒ송유섭

width. It may become a terrace, but limited by the setback from the due north. Be that as it may, it can be utilized as a chance to introduce formal aberration and outdoor space. As for *Quad*, the zigzag-patterned picturesque facades are applied and surrounded by a terrace. In a nutshell, the 4th floor is a terrace house.

The most eye-catching of *Quad* is the 5th floor. Its figure looks emphasized from afar like an abstract piece of sculpture put on a plinth. The penthouse is prepared to unfold unexpected indoor spaces. The top floor of a multi-family housing can become an exceptional part. It is also possible to design a building hovering in the sky, like the "castle in the sky" imagined surreally. The top floor is like a raw marble stone with its profile completed by the diagonal setback line. Architects can become sculptors, just like Michelangelo's phrase, "Every block of stone has a statue inside it and it is the task of the sculptor to discover it." [Figure 6]

The Chance Juxtaposition of a Sewing Machine and an Umbrella
on a Dissecting Table
After *Quad*, KimNam began to design different series of multi-family housing at the same time, based on a common type of the system but attempting to make variations case by case. They even experimented on structural variations. The recently completed multi-family housing in Ilwon-dong, *Warm and Cool*, pushes the limit of vanguardism that everyday architecture can reach.[Figure 7]

For the ground floor, a considerably long-span cantilever structure was radically applied to dispense with external columns. The vertical construction includes only the reinforced concrete bearing wall built narrowly in the midst and black-colored to hide its presence. The interior includes a neighborhood office, the internal floor of which was lowered to gain a ceiling height enough. The cantilever structure enabled transparent corner windows to be located deeply in the office, where one can enjoy the daylight pouring in.

The 2nd to 3rd floors are made up of a cantilever volume spreading areas as widely as possible. The internal floor plans combine residential studio units as in the Tetris game. The facades on which large square-shaped windows are punched regularly turn out to be the reinforced Vierendeel truss itself. This was the decision to prevent the floors from any possible deflection. Not latticed but gridded, as well as connected with not hinged but rigid joints, the Vierendeel truss is not without drawbacks in terms of structural mechanics but finds its special use in the architecture that values spatial efficacy. It has been rarely used in small-scale everyday architecture, but here is it justified for the convenience of ground-floor parking.

The building site of *Warm and Cool* is oriented to the due north so that a large area is under the influence of the setback

그 위에는 유리주택이 한쪽 끝단에 올려진다. 그 위로 다시 경사지붕 컨테이너가 적층된다. 하부에서 완벽한 수평을 이루던 양팔저울의 대칭성이 위기를 맞이하게 될 것이다. 상부의 과도한 비대칭성이 하부의 구조합리적 균형을 교란시킨다. 일상의 건축은 젠가놀이처럼 위태로운 구조적 도전에 직면한다. 로제르 카이요(Caillois)가 묘사했던 "진취적 생명성의 요소, 그러므로 위험과 모험"이 도입되고, 또 세실 발몽도(Balmond)의 표현처럼 "역동성이 개시"된 것이다.

이처럼 기묘한 형태와 구조적 관계는 건축 부지의 조건에 철저히 기반한 것이다. 건축의 내재적 가치에 대한 믿음을 간직한 채, 김남건축은 그 땅 위에 투영된 외재적 조건들과 '조용한 대화'를 계속한다. 각 세부 조건은 저마다 특수한 건축 해법을 요구한다. 건축 해법의 정당성 역시 외재적 조건과의 역학 관계를 통해 획득된다. 건축은 이러한 독자적 해법들이 자동 결합한 콜라주로서의 전체가 된다. 이렇게 건축의 내재적 가치는 외재적 조건과 공명한다. 사실 샬레도 그러했다. 각 부분들은 각자의 존재이유를 갖는다. 전체는 부분을 따를 뿐이다. 그렇게 낯선 형태가 알껍질을 깨고 태어난다. 생경함은 새로운 아름다움의 전제조건이다.

남성택

남성택은 한양대학교 건축학부 부교수이다. 서울대학교 건축학과를 졸업하고 프랑스 파리 마르느-라-발레 건축대학에서 석사 및 프랑스 공인 건축사를 획득했다. 이후 실무를 병행하며 스위스 로잔 연방공과대학(EPFL)에서 건축이론역사 제1연구실(LTH1) 소속으로 자크 뤼캉 교수의 지도 아래 '건축과 레디메이드'를 주제로 박사학위를 받았다. 2019년 뉴욕대학교, The Institute of Fine Arts 방문학자였다. 현재 건축을 중심으로 오브제 디자인에서 도시계획에 이르기까지, 즉 스케일의 구분 없이 삶에 관련된 인위적 환경에 관련된 구성, 구축, 변형 등 총체적 이론들과 디자인 연구에 관심을 두고 있다.

Sungtaeg Nam

Sungtaeg Nam is an Associate Professor at the School of Architecture, Hanyang University (Seoul Campus) since 2013. He holds Bachelor of Architecture from SNU (Korea), Diplome of Master and Architecte DPLG from EAV&T (France), and Ph.D. in EPFL (Switzerland), with thesis titled "The question of Readymade and its architectural appropriation". As architect, he has professional experience in Hankil Architect in Seoul, Du Besset et Lyon Architectes in Paris and PRS Architectes in Lausanne. In 2019, he was a visiting scholar at The Institute of Fine Arts, New York University (USA). The major focus of his research is on the various theories of Architecture, ranging from Object design (small) to Urbanism (big), without any distinction in scale. Fundamentally, he is interested in Composition, Construction, Space, Transformation, etc., related to human artificial environment.

regulation. Above the 4th floors, hence, only a narrow and long volume was buildable to the southern direction. The 4th and 5th floors constitute a penthouse for a family of four. Unexpectedly, a steel structure was used here in the name of saving space in a narrow volume. Actually, what is glimpsed here is the architects' hidden ambition and joy to experiment on structural complexity, just as they did in the chalet combining stone and wooden constructions. Surrounded by the northern terrace and the full-height glass window oriented to the roadside front, the 4th floor consists of one open space integrating the functions of living and dining. In contrast, the 5th floor can be encapsulated as a closed space, consisting of bedrooms with comparatively closed facades and finished with a simple gable roof. The whole structure is divided into diurnal and nocturnal spaces, the former for the family community and the latter for individual dwellers. Discovered here is a coexistence that is contrastive as well as complementary, just as the extreme transparency of Philip Johnson's Glass House could be complemented by the adjacent brick house's adamant opaqueness.

As a consequence, this alien form of the multi-family housing in Ilwon-dong is like Lautréamont's poetic line that became the Surrealists' official formula: "the chance juxtaposition of a sewing machine and an umbrella on a dissecting table".[1] The external form of the building appears a mystery. The Vierendeel truss put on the narrow black box spread long to both sides just like a seesaw, and on top of it, a glass house is put at one side. On top of it again, a gable-roofed container is accumulated. The symmetry of the balance scale forming a perfect horizontality at the lower part faces a crisis here. The excessive asymmetry at the upper part disturbs the structural-rationalist balance at the lower part. The everyday architecture faces a critical structural challenge as in the Jenga game. The "element of innovative vitality, therefore risk and adventure" is introduced as Roger Caillois described,[2] and "the dynamic is launched" as Cecil Balmond expressed.[3]

This peculiar form and structural relationship are based thoroughly on the building site's condition. Maintaining the belief in the immanent values of architecture, KimNam continues to make the "quiet conversation" with the extrinsic conditions projected over the site. Each specific condition requires a particular architectural solution case by case. This was true of the chalet. Each part has its own raison d'être. The whole just follows such parts. In this way, a strange form hatches out of the eggshell. The alienness is the precondition for a new beauty.

---

1 "Beau comme la rencontre fortuite sur une table de dissection d'une machine à coudre et d'un parapluie," in Lautréamont (Isidore Ducasse), Les chants de Maldoror, 1869.

2 Roger Caillois, "La dissymétrie", in Cohérences aventureuse, Paris, 1976, p. 256 : "élément de vitalité novatrice, donc risque et aventure".

3 Cecil Balmond, Informal, 2000, p. 27 : "La dynamique est lancée".

More Less Architects

모어레스 건축 | Youngsoo Kim 김영수

모어레스 건축사사무소는 급변하는 현시대에 조금은 더 명상적인 태도로 공간을 바라보고자 한다. 건축은 실용적인 물체로서 현시대의 기능을 수행하고 그 속에서 예술의 영향력을 찾고 감각의 공간들을 탐구해야 한다. 실용적인 건축 속에 무용한 가치와 낭만이 깃들어 새로운 경험의 공간으로 드러나길 바란다. "그래서 우리는 은근히 아름다운 공간을 좇고 있다."

김영수는 인하대학교와 인하대학원에서 공부하였으며 프랑스건축사회 11th 쟝프루베&김중업 Scholarship에 선발되었고 ㈜해안건축과 원오원 아키텍스, 프랑스 파리 DPA(Dominique Perrault Architecture) 등에서 다양한 규모의 프로젝트로 실무를 쌓았다. 현재 모어레스 건축사사무소 대표이며 인하대학교 건축학과 겸임교수로 출강 중이다.

In today's rapidly changing world, we desire to look at spaces with a more meditative perspective. Architecture, must perform the functions required in contemporary times with practical objects, but it should also find the influence of art out of them and explore the space of senses. It is our hope that practical architectural works will emerge as spaces of new experiences that are infused with impractical values and romanticism. "That's why we are subtly pursuing beautiful spaces."

Youngsoo Kim studied at Inha University and its Graduate School. He was selected as a recipient of the 11th Jean-Prouvé & Kim Chung-up Scholarship by the French Architectural Society, and has worked on projects of various scales at Haeahn Architecture Co., Ltd. and One O One Architects in Korea and Dominique Perrault Architecture (DPA) in Paris, France. He is currently the principal of Moreless Architects and an adjunct professor at the Department of Architecture, Inha University.

Impractical Interests

무용無用한 관심사

재료 본연의 성질, 건축 요소의 자리, 사물과 공간의 관계 등 건축 본질에 대한 집요한 탐색과 사유 과정이 건축가 개인의 지적 감수성(intellectual sensitivity)에서 비롯된 태도를 넘어 젊은 건축가상의 당대성과 미래적 역할에 방향타가 될 만한 유의미한 모습으로 전달됐다. 심사를 하면서 과연 이들에게 건축이 고통스럽지 않고 즐겁다면 지속가능한 내적 영역이 강건하게 있는가를 추측해 보았고, 대표적으로 김영수가 그랬다. 엄격하게 조정한 공간의 형태, 재료의 두께, 빛의 강도를 현장에서 섬세하게 통솔하여 완성도 높은 건축물로 실현해 가는 과정 속에 건축가 특유의 오기와 환희 그리고 스스로를 향한 비평이 즐겁고 균형 있게 자리 잡혀 있음이 엿보였다.

― 심사평 중

His persistent exploration and reasoning process on the essence of architecture, such as the inherent properties of materials, the placement of architectural elements, and the relationship between objects and space, was so significantly conveyed as to become a rudder that could guide the contemporaneity and future role of the KYAA beyond the architect's individual intellectual sensitivity. During the jury process, I speculated whether these architects – if architecture is not distressing but enjoyable for them – have an inside strong enough to be sustainable, and found its exemplar from Youngsoo Kim. In the process of strictly adjusting the form of spaces, the thickness of materials, and the intensity of light and meticulously orchestrating them on site into a building in great perfection, he revealed a delightful and balanced presence of the architect's peculiar passion, joy, and self-critique.

― From Jury's comment

- 도락道樂의 마음 140
- 상상과 확신 136
- 부분에서 전체 122
- 사유의 노력 112
- 건축의 시작 102
- 수의 질서와 호기심 94
- 질서의 아름다움과 분위기 88

The Beauty of Order and Atmosphere	*89*
Numerical Order and Curiosity	*95*
Beginning of Architecture	*103*
Efforts of Reasoning	*113*
From Parts to Whole	*123*
Imagination and Confidence	*137*
Heart of Amusement	*141*

More Less Architects

질서의 아름다움과 분위기

나지요네
Nasilonner
ⓒTaxu Lee

나는 낯선 분위기에 강한 호기심을 느꼈다. 그는 어떤 내면을 가졌기에 이 같은 피상적인 공간을 만들어냈는가? 당시 난 육체적 허기를 정신적 허기로 착각하여 그 복도에 빠져든 것이리라.

　　광화문 근처 국밥집이었다. 그곳에는 아주 독특한 구조를 가진 복도가 있다. 식사를 하는 넓은 홀은 층고가 높아 사람들의 이야기가 적당히 울리는, 괜찮은 분위기의 공간이다. 하지만 화장실로 연결된 복도는 꽤 긴 형태로 한 사람이 겨우 지나갈 만한 폭에 천장은 손에 닿았다. 흐릿한 조명마저 없었다면 이 복도의 끝은 긴 어둠의 나락으로 이어져 사라져 버릴 것만 같은 느낌을 주었을 것이다. 그 깊은 복도의 끝에 이르러 더듬더듬 돌아서면 운명의 화장실에 다다른다.

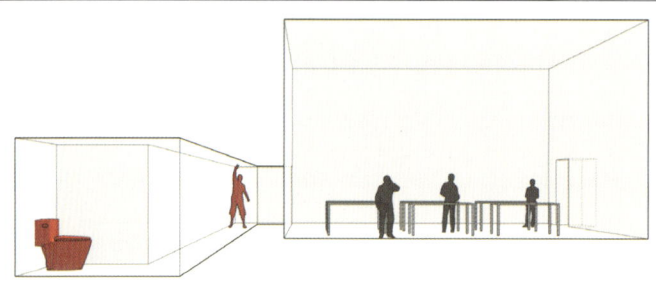

restaurant corridor

The Beauty of Order and Atmosphere

The unusual atmosphere aroused strong curiosity in me. What spirit on earth created such a superficial space? Perhaps, I had mistaken my physical hunger for mental hunger and got pulled into the hallway.

It was a gukbap restaurant nearby Gwanghwamun. What was unique about the place was the hallway. The dining area was spacious with a high ceiling, creating a comfortable atmosphere with decent noise levels. But the lengthy hallway to the restroom was barely wide enough for one person to pass through, not to mention the ceiling was so low that you could reach with your hand. Without the dim lighting that was the only source of light, it would have felt as if the end of the hallway would disappear into a long, dark abyss. When you manage to make it to the end of the deep hallway and feel your way around, you will arrive at the destined restroom.

| More Less Architects

의도치 않게 만들어진 공간이 독특한 분위기를 자아낼 때가 있다. 그런 공간을 만나면 나는 설렌다. 때로는 공간이 긴장감을 만들고, 때로는 공간이 눈치를 보게 한다. 또 좋은 공간은 내게 평온함을 준다.

누구에게나 선호하는 분위기 있는 공간이 있다. 날씨가 좋은 날이면 가끔 커피를 마시는 장소가 있다. 사무실 건너 옛 공간의 신사옥 1층 카페는 구(舊)사옥과 신(新)사옥 그리고 한옥 사이에 아늑한 마당을 품고 있다. 적당한 크기의 마당, 단차가 있는 바닥, 벽돌과 한옥, 커튼월 같은 재료, 거기에 담쟁이넝쿨까지, 그것들이 만드는 하나의 풍경이 있다. 이러한 분위기 속에서 나는 커피를, 아니 공간을 즐긴다. 기능과 편의를 넘어 분위기로서 존재하는 공간은 나에게 평온함을 준다.

cafe

건축은 물리적 공간을 만드는 실용적인 행위이다. 하지만 단순히 물리적 공간으로만 정의되지 않으며 실용적 행위로만 구축되지도 않는다. 건축은 건축가의 생각과 감각이 더해져 공간의 분위기로 나타나게 된다. 그런 분위기는 공간뿐 아니라 형태, 재료, 사람, 자연의 일부 등이 조합되어 만들어진다. 공간 분위기는 건축가의 타고난 솜씨로 담아내든, 취향의 참조를 통해 담아내든, 거창한 이론과 논리로 담아내든 물리적 공간을 넘어 각자의 예술이 된다. 그 예술은 실용적인 기능으로, 편리함으로, 화려한 형태로 보여지기도 하고 잘 인식되지 않는 분위기로 나타나기도 한다. 모든 건축예술에 의미가 있지만, 나는 실용성과 먼, 각자의 방식으로 드러나는 건축의 공간감과 분위기에 더 큰 호기심이 있다.

루민연남
Lumen Yeonnam
ⓒJoel Moritz

Impractical Interests

Among so many buildings around us, there are times when an unintentionally created space gives off a unique atmosphere. It is those spaces that excite me. At times, there are spaces that make me nervous while some make me self-conscious. Then there are good spaces that bring me peace.

We all have a preferred space with a particular atmosphere. There is a place I go for coffee when the weather is nice. It is a coffee shop on the first floor of the new building, which is located across from my office. The cozy yard in front of the place is surrounded by an old building, a new building, and Hanok. A decent-sized yard, split levels of floors, materials like bricks, Hanok, and curtain walls, and even the vines on the fence create a landscape of its own. In this atmosphere, I enjoy my coffee, or rather, the space. This is a space that brings me peace, existing as an atmosphere beyond functionality and convenience.

Architecture is the practical act of creating physical space. Nonetheless, it is not simply limited to physical space or constructed solely by practical actions. Architecture is expressed as the atmosphere of a space with a touch of the architect's thoughts and senses, and such an atmosphere is created by combining forms, materials, people, and part of nature. This spatial atmosphere becomes one's art, going beyond just physical space, whether it is captured by the architect's innate skills, through a referenced taste or a grandiose theory or logic. The art may appear to be practically functional, convenient, or ornately formed or as an ambience not easily recognized. While every architectural art is meaningful, I am particularly interested in the spatial sense and atmosphere of architecture that is expressed in its own way, far from practicality.

모어레스 건축 | More Less Architects

좋은 공간은 건축의 실용성, 형태, 논리를 넘어 그저 분위기로 나를 이끈다. 그리고 설득하지도, 강요하지도 않으며 나를 도취시킨다. 그렇게 공간은 숭고한 건축이 된다. 그런 공간은 경험을 해도 모든 감각과 의도를 지레짐작하여 정의할 수 없으며 도면이라는 형식으로도 완전히 이해할 수 없다.

그렇기에 무용한 관심사 안에서 공간 미학의 세계를 찾고 경험하기를 지속해 간다. 우리는 하나씩 직접 보고 만지고 느끼며 나아가고 있다고 생각한다. 그렇게 쌓인 공간은 하나하나 우리의 감각 속에 각인되어 조금씩 의도하는 공간에 확신을 더해간다. 그 과정은 나에게 스스로 공간의 가치를 부여하는 것이다. 그리하여 나는 여전히 돌아다니고 두리번거리며 공간 미학의 조각을 사유해 간다.

도산 알로하
Dosan Aloha
ⓒJoel Moritz

Impractical Interests

A good space goes beyond the practicality, form, and logic of architecture and simply draws me into its atmosphere. It doesn't persuade me or force me; it just intoxicates me. That's how a space becomes sublime architecture. Even if we experience such a space, we cannot hastily guess and define all its senses and intentions or even fully understand them in the form of a drawing.

This is why I constantly seek out and experience the world of spatial aesthetics within impractical interests. I believe we move forward by seeing, touching, and feeling at first hand. The spaces accumulated this way become engraved in our senses one by one, gradually adding confidence to the intended space. I consider this to be a process of granting the value of space to myself. This is why I am always wandering around and looking around, contemplating pieces of spatial aesthetics.

수의 질서와 호기심

'$x^n + y^n = z^n$(n은 3 이상의 정수)을 만족하는 정수해 x, y, z는 존재하지 않는다.'
17세기 수학자 페르마(Pierre de Ferma)는 마지막 정리를 남기고 증명도 없이 세상을 떠났다. 페르마의 마지막 정리가 증명되기까지 천재 수학자들은 350년의 시간을 보냈다. 마지막 정리는 어떠한 실용적 의미가 있는지 이해되지 않으나 수학자들은 진실에 대한 호기심과 그 과정의 가치를 알기에 오랜 시간을 보냈다. 오랜 수고를 들인 증명은 단지 수에 대한 진리였지만 진리는 그 자체로 아름답다.

가온누리
Gaon-Ruri
ⓒJoel Moritz

대수학은 순수학문으로서 실용적인 가치를 논하지 않는다. 수학의 정리는 증명의 과정을 통해 또 다른 수학의 증명으로 응용되고 새로운 정리가 만들어진다. 그렇게 증명된 진리는 다시 수학의 근본을 이루는 기초가 되고 그 위에 새로운 기초가 쌓여 다양한 학문에서 응용되고 실용화되는 것이다.

수의 정리는 증명되어 학문의 진리가 된다. 과학의 탐구는 실험을 통한 진실에 '가까운' 이론에 이른다. 건축은 진리나 진실에 가까울 수 없다 해도 모두가 공감할 만한 공간감은 정의될 수 있지 않을까. 미적인 판단은 주관적이지만 보편 타당성 또한 가질 수 있다고 하지 않는가? 이성적 보편성뿐만 아니라 감성적 보편성 또한 논의될 수

Numerical Order and Curiosity

"No three positive integers x, y, and z satisfy the equation $x^n+y^n=z^n$ (for any integer value of n greater than 2)."

Without proving his "Last Theorem," a leading mathematician of the 17th century, Pierre de Fermat, passed away. It took more than three centuries for prominent mathematicians to prove Fermat's last theorem. While it's hard to fathom the practical significance behind the last theorem, mathematicians spent endless hours in search of the truth, knowing the value of the process. Even though the long-time efforts were only to prove a numerical truth, the truth is beautiful in itself.

Algebra is not about practical value since it is a pure science. Through the process of proving, a mathematical theorem is applied to another mathematical proof, creating a new theorem. The truth proven this way becomes the foundation of mathematics, on top of which new foundations are built to be applied and put to practical use in various disciplines.

Theorems on numbers become the academic truth when they are proven. Scientific inquiries lead to theories that are "closer" to the truth through experiments. Even if architecture cannot be close to the truth or verity, perhaps its sense of space can be defined as that in which everyone can feel sympathy. Isn't it said that aesthetic judgments are subjective but can also

있다고 한다. 그렇다면 개인적 취향을 넘어 모두가 공감할 수 있는 보편 타당한 감성적 공간예술이 있지 않을까?

르 코르뷔지에(Le Corbusier)의 모듈러(Le Modulor)를 보며 그의 호기심과 탐구, 열정에 탄복한 때가 있었다. 돌이켜보면 그가 찾았던 것은 보편 타당한 건축 공간예술이 아니었을까? 그는 모듈러에 대해 '인체의 치수와 수학의 결합에서 만들어지는 것을 계량하는 도구이자 다양한 디자인에도 적용할 수 있는 척도'라고 설명했다. 공간개념에 대한 의의를 탐구하면서도 공작물의 형태를 규정하고 있는 척도에 관심을 둔 비례체계이다.

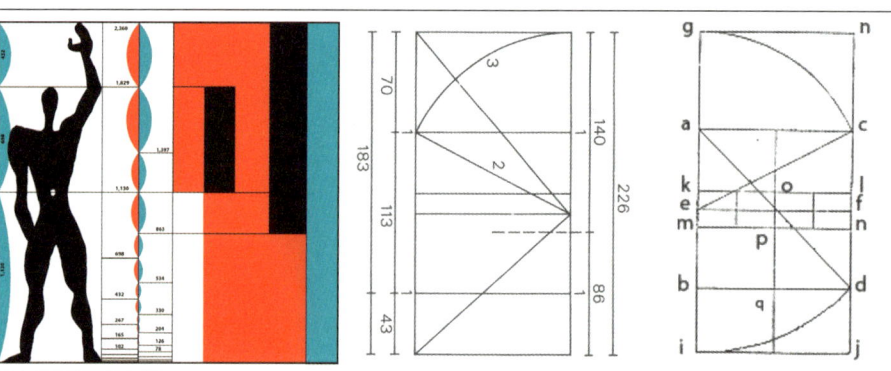

Le Modulor

우리가 잘 아는 모듈러의 이미지는 단순히 정리된 다이어그램 같지만, 사실은 수와 기하학을 이용한 정리에 신체의 비를 결합시켜 수열(27, 43, 70, 113, 183, 226)을 만든 것이다. 르 코르뷔지에는 그 비례의 가치를 입증하려고 다양한 사원(살라의 수도원, 이집트 신전의 부조, 소피아성당, 폼페이 포럼 등)을 쫓아다니며 검증해 나갔다. 내게는 모듈러라는 결과물을 떠나 그가 찾으려는 미적 판단의 호기심과 수와 비례에 관한 탐구 과정이 아름다움이다.

대학원 시절 루돌프 쉰들러(Rudolph M. Schindler)의 건축을 오래 연구한 은사님으로부터 쉰들러의 건축을 알았다. 쉰들러의 건축은 엄격한 모듈 속에 건축의 요소들을 구성한다. 평면과 입면은 모듈 체계를 가지며 각각의 공간은 정수의 비례 관계를 갖는다. 그의 건축은 단지 공간 구축법에 머무르지 않고 자신만의 방식으로 공간을 조합 배열하며 시퀀스와 분위기까지 이끌어간다. 알바로 시자(Alvaro Siza)는 어느 글에서 쉰들러의 주택을 설명할 수 없는 '편안함'이라 말했다. "그런 곳에 들어갈 때의 평화로움과 행복함은 거의 천국과 같은 느낌이었다"고 찬사했는데 도대체 무엇이 시자의 마음을 그토록 설레게 했던 것일까? 단순히 모듈과 공간의 비례만은 아니겠으나 그 또한 주요한 요소 중 하나였을 것이다.

have universal validity? This refers not only to rational universality but also to emotional universality. Then, wouldn't there be an emotional spatial art that is universally valid beyond one's personal taste in which everyone can feel sympathy?

I recall seeing Le Corbusier's Le Modulor and admiring his curiosity, exploration, and passion. Looking back, wasn't it a universally valid architectural spatial art that he sought for? He explained the Modulor as "a range of harmonious measurements to suit the human scale, universally applicable to architecture and to mechanical things." It is a proportional system that explores the meaning of the concept of space while paying attention to the scale that defines the form of a structure.

The image of the Modulor that we are familiar with looks like a simple diagram, but in fact, it is a sequence (27, 43, 70, 113, 183, and 226) created by combining the proportions of the human body, using numbers and geometry. To prove the value of the proportions, Le Corbusier went after and verified various temples (Sala monastery, reliefs of Egyptian temples, Hagia Sophia, Pompeii Forum, etc.). To me, beauty lies in his curiosity for aesthetic judgments and the process of exploring numbers and proportions rather than the Modulor itself.

It was in graduate school that I first learned about Rudolph M. Schindler from my professor who had studied Schindler's architecture for a long time. His architecture organizes architectural elements in strict modules: the floor plans and elevations are based on a modular system, and each space has proportional relationships. His architecture is not just limited to spatial construction but also combines and arranges spaces in his own way to create sequences and atmosphere. Álvaro Siza, once in an article, described Schindler's house as an indescribable "comfort." He praised the house by saying, "The peace and happiness from entering such a place almost felt like heaven." What could have possibly made Siza so sentimental? It may not be simply the proportions of modules and spaces but those could have been one of the major factors.

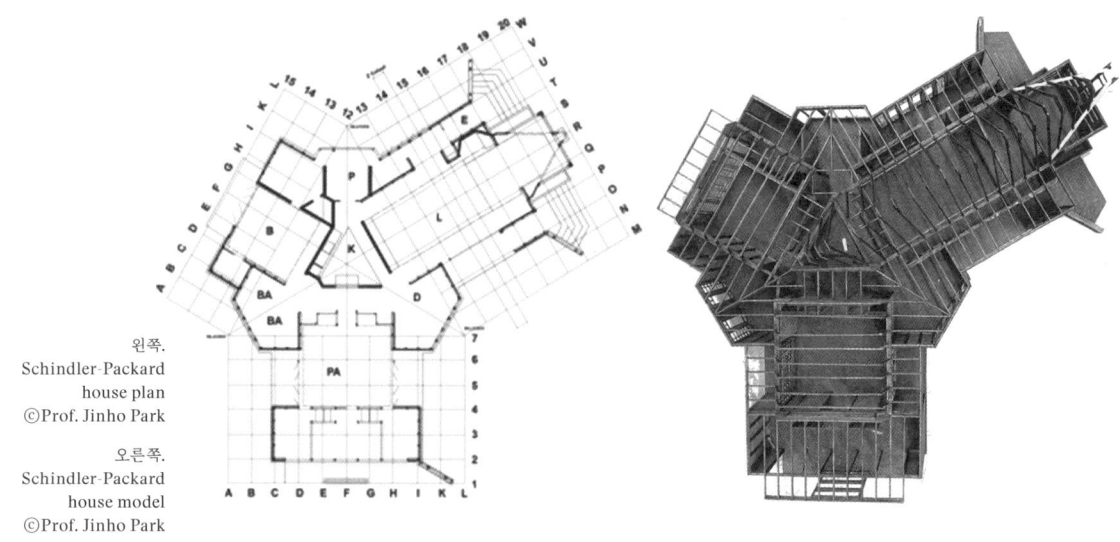

왼쪽.
Schindler-Packard
house plan
ⓒProf. Jinho Park

오른쪽.
Schindler-Packard
house model
ⓒProf. Jinho Park

주세페 테라니(Giuseppe Terragni)의 건축은 고전주의적 합리주의 비례와 수학, 이미 존재하여 주어지는 질서를 전제로 한다. 그는 새로운 건축에서 논리와 합리성을 고수하는 방법을 취하지만 자신만의 방식으로 풀어낸다. 그만의 방식으로 풀어낸 질서의 구축이 공간의 힘이자 형태의 아름다움을 정의하는 주요한 요소이지 않을까 생각한다.

안드레아 팔라디오(Andrea Palladio)는 방을 디자인할 때 일곱 가지 아름답고 조화로운 비율을 제시했다. 어설픈 지식으로 너무 멀리 온 듯하지만, 어쨌든 내가 관심을 가진 건축가(루이스 칸, 르 코르뷔지에, 주세페 테라니, 루돌프 쉰들러, 안드레아 팔라디오 등)의 작업에는 겉으로 드러나지 않아도 공간 속에 수와 관련된 질서와 비례가 각자의 방식으로 담겨 있다.

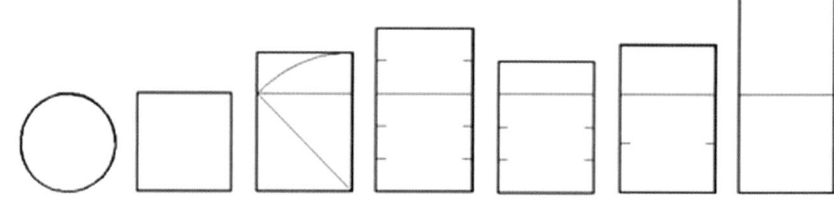

7 Ideal Plan Shapes
For Rooms
by Andrea Palladio

건축에 대한 우리의 관심사는 결국 수와 관련된 고리타분한 고전적 질서 속에 머물러 있는지도 모른다. 단순히 비례를 통해 우주의 질서와 아름다움을 찾으려는 것은 아니고, 그런 시대가 지난 지도 오래다. 그저 공간과 형태 속에 의미를 담지 않는 기본적인 미학적 가치로 남겨두고 싶은 것뿐이다. 그렇기에 절대적인 가치라기보다 부분적인 생각이며, 기본적인 방법론이 되는 것이다. 수의 질서와 배열에 대한 고민은 공간의 전개와 형태, 볼륨의 기준이 필요한 순간 나만의 방식으로 취하는 것이다.

앞서 말했듯 다양한 공간을 경험해보면 건축은 단순히 비례만으로 공간감을 형성하거나 감동을 주지 않는다는 것을 알 수 있다. 건축에는 고려해야 할 요소도, 변수도 많고 다양하다. 르 코르뷔지에 역시 롱샹 성당을 계획하며 "상상을 방해하거나 사물의 절대성을 주장하거나 혹은 발명을 움츠리게 하는 경우에는 원칙적으로

Giuseppe Terragni's architecture is premised on classical rationalism, proportions, mathematics, and pre-existing orders. He takes a strict adherence to logic and rationality in new architecture but solves it in his own way. I believe the construction of order, expressed in his own way, could create the power of space as a major element that defines the beauty of form.

When Andrea Palladio designed a room, he suggested seven sets of the most beautiful and harmonious proportions. It seems I have gone too far with my limited knowledge; anyway, in the works of the architects I am interested in (Louis Kahn, Le Corbusier, Giuseppe Terragni, Rudolf Schindler, Andrea Palladio, etc.), orders and proportions may not seem apparent in appearance but exist in relevance to numbers within each space in different ways.

Our interest in architecture may ultimately remain in the old-fashioned classical order related to numbers. It is not simply an attempt to find order and beauty in the universe through proportions, and we are no longer in that era. I just want to leave it as a basic aesthetic value that does not necessarily put meaning in space and form. As such, it is a partial idea, a basic methodology, rather than an absolute value. I will handle the concerns about the order and arrangement of numbers in my own way when the time comes for me to need the standards of spatial planning, form, and volume.

As mentioned earlier, experiences in various spaces will teach you that architecture does not create a sense of space or move people's hearts simply by proportions. In architecture, there are numerous and diverse elements and variables to consider. Le Corbusier, when designing the Ronchamp chapel, said, "I am opposed to the 'Modulor' if it interferes with imagination,

'모듈러'를 반대한다. 상호관계의 절대적 속성을 믿는다. 그리고 상호관계는 그 정의에서도 가변적이며 잡다하며 무수하다"고 말했다.

불확실하고 다원적인 현시대에 우리의 무용한 관심사는 형식의 질서를 만들고 이를 바탕으로 건축의 의미를 생성할 또 다른 가치를 사유해 나가려고 한다. 그렇게 우리가 찾는 것은 여전히 짙은 안개 속에 있다.

루민연남
Lumen Yeonnam
ⓒJoel Moritz

insists on the absoluteness of objects, or discourages inventions. I believe in the absolute nature of interrelations. And interrelations, in their definitions, are variable, miscellaneous, and innumerable."

In this era of uncertainty and pluralism, our impractical interests attempt to create an order of form and based on that, try to think of another value that will create the meaning of architecture. So what we are in search of still remains in a thick fog.

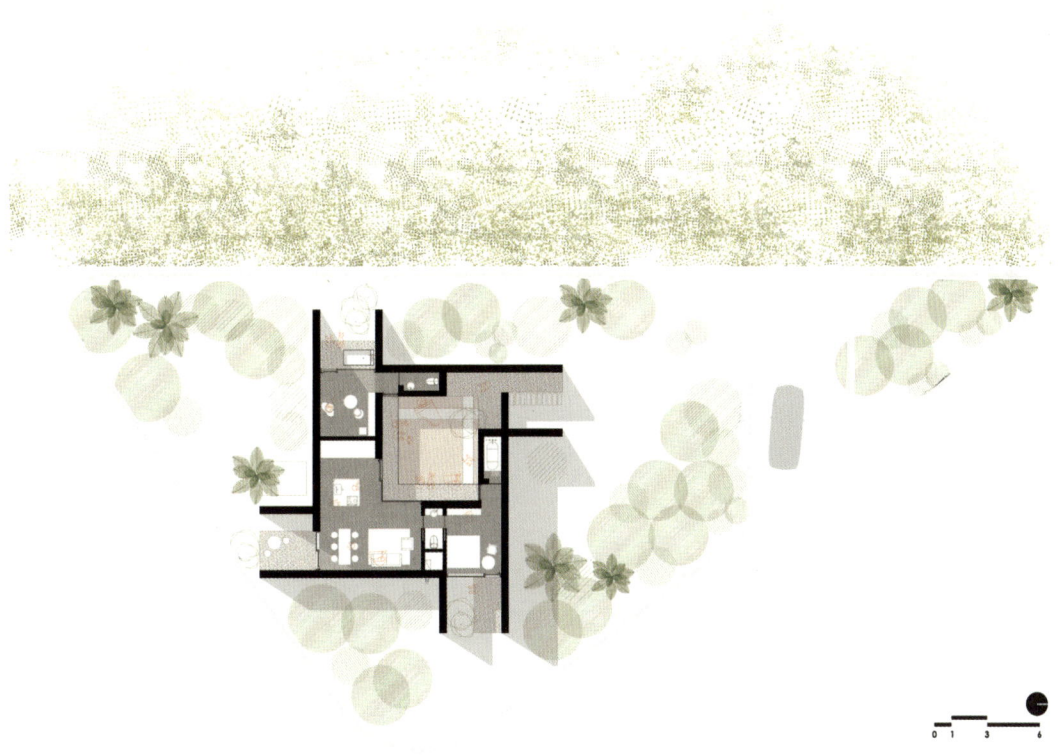

모어레스 | More Less Architects

건축의 시작

나지요네
Nasilonner
ⓒTaxu Lee

Beginning of Architecture

More Less Architects

2018년 준공한 <나지요네(Nasilonner)>는 공식적인 첫 프로젝트이자 유일한 프로젝트였기에 욕망이 가득하다. 초기 설계는 어수선했고 '개념'이란 집착 속에 계획안은 땅속에 묻혔다가 땅 위로 솟았다가 하며 심의, 허가, 입찰을 반복했다. 열정 가득한 욕망과 기대는 무너졌고 물리적 한계와 체념을 통해 상황은 평온해졌다. 그렇게 나는 현실과 마주했다.

클라이언트는 건축가의 열정을 수용했으나 예산 초과는 수용하지 못했다. 시공자와 그 외 모든 협력관계는 이전의 안락함과 다른, 격렬한 현장의 전쟁터였다. 너무나 당연한 이치였지만 스스로 전문가라 생각했던 10년의 경력은 나의 오만함이었고, 이전과 다른 시장에서 나의 건축이 작동하기에는 또 다른 경험의 임계량이 필요했다. 그러한 시간 속에서 생각도, 태도도 조금은 더 다듬어진다. 슬픔과 체념도 풍화되고 주어진 조건과 한계를 인식하며 할 수 있는, 또 해야 하는 건축을 하게 된다.

내게 남겨진 것은 여유로운 면적도, 고급스러운 재료도, 섬세한 디테일도 아니었다. 그것은 공간의 구성, 구축의 질서뿐이었다. 최소한 질서부터 건축 공간 속에 담아낼 수 있기를 바랐고 또 그것이면 충분했다.

땅의 콘텍스트는 내부공간만큼 외부공간을 중요하게 만들었다. 중정을 중심으로 공간을 연결하고, 각각의 공간은 다시 독립된 외부공간으로 전개하며 내외부의 관계를 만든다. 공간의 관계가 명확한 질서 위에 드러나길 바랐다. 공간은 회전을 통한 대칭변환으로 평면을 구성하고 기준공간과 확장공간, 부속공간의 비율 (1a : 2a : 0.5a X 1a)을 다루며 배열을 통해 질서를 부여했다.

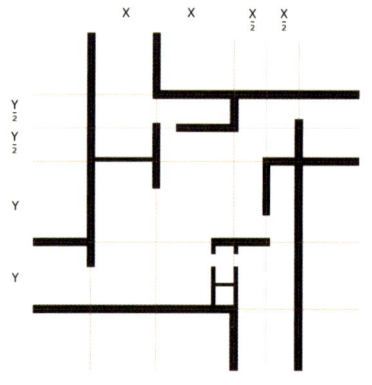 

공간의 관계, 관계 비, 배열 등 질서를 부여하는 것에 의미를 뒀다. 이러한 질서는 공간의 관계 속에서 개별적 공간감을 만드는 중요한 부분이 되며, 또 건축 전체를 이해하는 방법이기도 하다. 각 공간의 규모와 관계는 의도한 흐름 속에서 구성되고 거기에 형태와 구조가 어울려 건축이 된다.

나는 독특한 형태나 자유로운 선의 아름다움보다 질서의 아름다움이 좋다.

Nasillonner, which was completed in 2018, was full of ambition as it was the first and only official project. The design was out of place at the early phase, and being obsessed with a "concept," it went underground or above ground through the repeated project cycle of deliberation, approval, and bidding. In the end, ambitious desires and expectations were shattered, and I found peace through physical limitations and resignation. I was face to face with reality.

The client accepted the architect's passion, but did not embrace going over budget. Working with the contractor and all other partners was like a fierce battlefield, far from the previous comforts. This was indeed an obvious flow of practice, but the ten years of experience I gave myself credit for was my arrogance. For my building to survive in this unfamiliar market, I needed to accumulate a critical mass of experience. And the process will help me a little to refine my thoughts and attitude. Disappointments and resignation will weather away, and architecture becomes what I can do and must do under given conditions and limitations.

What I was left with was not a spacious area, luxurious materials, or specific details. It was only the composition of space, the order of construction. I hoped that I could at least put order in the architectural space, and that was going to be enough for me.

The site context made me pay attention to the exterior as much as the interior. Around the courtyard, all the spaces were connected, and yet, each space was an external space of its own, creating a relationship between the inside and outside. I wanted these spatial relationships to be visible in a clear order. The floor plans were designed in rotational symmetry, and order was given by arranging the spaces (reference space, extended space, and annex space) in the order of a ratio (1a:2a:0.5ax1a).

Consideration was given to the order of spaces, such as spatial relations, relationship ratios, arrangement, etc. This is an important part of creating an individual sense of space within spatial relations and is also a way to understand the entire building. The scale of each space and its relation with other spaces are organized within an intended flow, and when forms and structures are added, they become architecture.

I prefer the beauty of order rather than the beauty of unique forms or free lines.

공간은 역보 구조와 뻗어 나온 날개 벽을 통해 의도한 기능과 형태를 이룬다. 역보 구조를 통해 공간은 노출 콘크리트 천장을 드러내고, 뻗어 나온 벽체는 내부공간과 같은 비율의 외부공간을 만든다. 내외부의 관계로 만들어진 공간 비율, 연속된 벽과 바닥의(콘크리트 벽체) 재료는 공간의 연속성을 나타낸다.

나지요네
Nasilonner
ⓒTaxu Lee

질서의 구축 속에 무용한 상상의 의도를 더한다. 뻗어 나온 날개 벽은 구조적 기능 이상의 두께(400mm)를 갖는다. 솟아오른 역보는 구조적 기능 이상의 보춤(600mm)을 갖는다. 툇마루의 두께(400mm)도 기능 이상의 크기가 된다. 보가 가진, 벽이 가진, 또 툇마루가 가진 실용적 기능 이상의 상상들이다. 쉽게 인식되지 않는 요소이자 무용한 기능이지만, 그것은 우리가 상상한 분위기를 만든다.

<나지요네>를 마무리하며 구축적 질서를 통한 공간 구성, 배열에 대한 생각은 확장되어갔다. 건축의 양과 치수의 비례 관계를 다루는 생각(symetria)이다. <나지요네>가 공간의 기하학적 대칭 변환으로서 회전, 반사, 병진(竝進)의 기본적인 요소를 바탕으로 한 평면 구성 중에 회전이었다면, <사라한남(Sara hannam)>은 반사의 대칭변환을 통한 구성이다. 도심 속 법적 규제가 가득한 작은 대지에 고밀도 공간을 계획하면서 질서를 바탕으로 간결하게 구축하려 했던 프로젝트이다.

Impractical Interests

A space achieves its intended function and form through the inverted beams and extended wing walls. The inverted beams allow the space to reveal an exposed concrete ceiling, and the extended walls create an external space all proportional to the internal space. The spatial ratio created by the relation

between the interior and exterior and the materials of the continuous wall and floor (concrete) indicate the continuity of space.

I then add impractical, imagined intentions to the establishment of order. The extended wing wall has a thickness (400 mm) that is beyond its structural function. The raised beam (600 mm) has more than just a structural function. The thickness of toenmaru (400 mm) also exceeds its function. These are imaginations that go beyond the practical functions of beams, walls, and toenmaru. While there are elements not easily recognized and have impractical functions, they create the atmosphere we had in mind.

Coming to the completion of *Nasillonner*, my thoughts on spatial composition and arrangements through constructional order began to expand. This was an idea (symetria) on the proportional relationship between quantity and dimensions in architecture. Among the planar compositions

사라한남
Sara Hannam
ⓒJinbo Choi

소규모 1~2인 사무실의 기능을 담당할 유니트 형태의 공간은 코어를 중심으로 대칭변환의 반사를 기본으로 한다. 이 유니트는 법적인 조건 아래 증식되거나 사라지는 데 이로써 전체 공간을 이루고자 했다. 최대 건축면적, 최대 용적률, 일조사선 내 최대 높이, 최소한의 주차공간, 구조의 배열 등 복잡한 조건들이 질서 안에서 만족하고 내부 기둥, 코너구조, 외부 기둥열, 커튼월의 모듈은 정리되어 간결한 공간을 이루게 된다. 복잡한 산식 속에 간단한 해답과 같은 평면은 나에겐 아름다움이다.

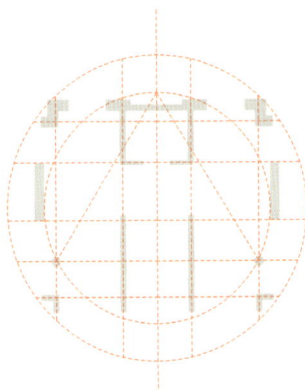

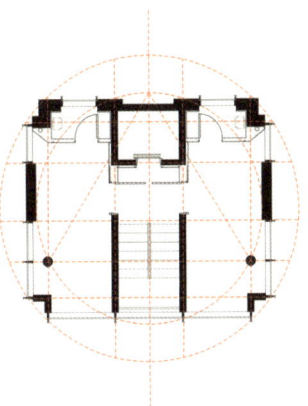

based on the basic elements of rotation, reflection, and translation as the geometrical symmetry of space, *Nasillonner* would be categorized under rotation while *Sara Hannam* would be a composition through reflection symmetry. Building a high-density space on a small site full of legal regulations of the city, we tried to keep it as simple as we can on the foundation of order.

    The unit-type space that will function as a small office for one to two persons is based on reflection symmetry around the core. We wanted to create an overall space using these units, which can increase or disappear under legal conditions. Complex conditions, such as the maximum building area, the maximum floor area ratio, the maximum height within the diagonal setback line, the minimum amount of parking space, and the arrangement of structures, were all satisfied within order, and the modules of internal pillars, corner structures, external pillar rows, and curtain walls were arranged to create a simple space. To me, a floor plan that seems like a simple solution in a complex arithmetic is a beauty.

사라한남
Sara Hannam
ⓒJoel Moritz

공간의 간결함만큼 재료의 간결함을 원했다. 주 마감재료는 콘크리트와 유리, 금속(아연도패널)의 조합으로 최소한의 재료로 제안했고 고객은 '콘크리트의 러프함을 좋아한다'고 말했다. 고객이 재료가 가진 물성 혹은 그 특성을 이해하고 공감해준다는 것은 몰입의 동기가 된다. 그렇게 우리는 각각의 재료가 스스로의 물성을 마음껏 드러내는 방식으로 이끌어갔다. 콘크리트는 거푸집이 만들어낸 투박하고 거친 표면을 보여주고 아연도패널은 벽체와 천장의 최종 마감재가 되어 거친 콘크리트 사이에서 금속의 섬세함을 드러낸다. 유리는 프레임을 감추고 투명함으로 온전히 공간을 드러내는 방식을 취한다.

우리는 공간도 재료도 그렇게 순순한 아름다움이 되길 바랐다.

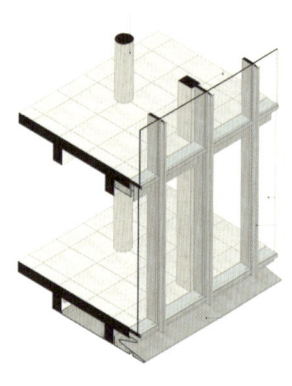

Exposed Concrete
Porcelain tile

Low-E insulated glass unit
Aluminium injection bar

A galvanized steel sheet

사라한남
Sara Hannam
ⓒJinbo Choi

We wanted to keep simplicity in not just space but also materials. We proposed that the main finishing material will be a combination of concrete, glass, and metal (galvanized steel sheets), keeping materials to a minimum, and the client indicated their "preference for the roughness of concrete." When a client understands and connects with the physical properties or characteristics of a material, it serves as a motive for our immersion. And so we allowed each material to freely reveal its own physical properties. Concrete showed off its rough and coarse surface created by formwork, and galvanized steel sheets became the finishing material for walls and ceilings, revealing the delicacy of metal within the rough concrete. As for glass, we hid the frames so that spatial transparency could fully be revealed.

We wanted the space, as well as the materials, to become a pure beauty.

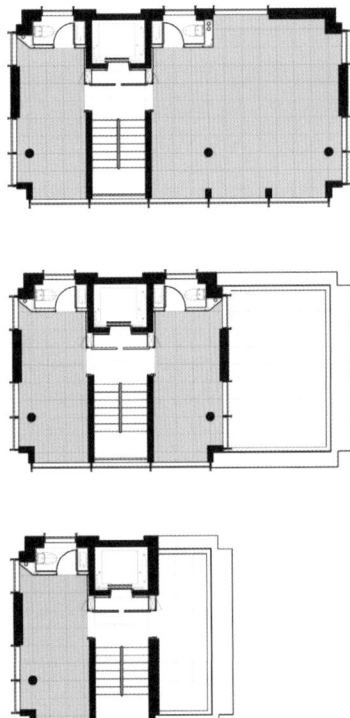

무용無用한 관심사

More Less Architects

사유의 노력

건축가는 열성(劣性)으로서 당당히 윤리와 가치를 주장하고 온전히 자신의 철학을 클라이언트에게 강요할 수 있는 존재가 아니다. 열등한 사회적 위치에서 거창한 생각들을 구구절절 늘어놓고 클라이언트를 설득하며 건축의 가치를 놓치지 않으려고 한다. 그 속에서 벌어지는 우리의 타협과 체념이 건축가로서 진보라고 할 수는 없으나 적어도 현시대를 극복해 나가려는 노력이라 생각한다. 우리는 그렇게 한걸음씩 나아간다.

우리는 시대와 문화, 사물과 현상을 바라보고 개념을 세우며 스스로 이야기를 정리해 가는 일을 즐겁게 수행해야 한다. 그것이 철학적 의미를 갖거나 현시대를 통찰하는 유의미한 일은 아닐지언정 그 나름의 사소한 생각이라도 이어가며 사유하는 노력이 계속 되어야 하는 것이다. 클라이언트가 원하지 않는 노력이며 쓸데없는 의미 부여라 말할 수도 있지만 그것의 가치는 고민하는 자만이 스스로 알 것이다.

도산 알로하
Dosan Aloha
ⓒJoel Moritz

상업지구 보행가로의 건물들은 도시구조, 사회, 문화 등 다양한 흐름과 함께 갑작스러운 변화를 맞이하곤 한다. 설계 당시의 의도와 상관없이 상권 변화에 따른 새로운 쓰임으로, 또 새로운 분위기로 변화가 요구된다. 대표적인 상업지구인 압구정에 위치한 <도산 알로하(Dosan Aloha)> 프로젝트 역시 이러한 흐름 속에서 진행되었다.

Efforts of Reasoning

Playing the recessive role, architects cannot firmly assert ethics and values and completely force their own philosophy on clients. Architects in the inferior social position express grandiose ideas and persuade clients all the while not letting go of the value of architecture. While the compromises and resignations occurring in the process may not be considered progress, I believe they at least represent an effort to navigate the challenges of the current era. This is how we move forward, one step at a time.

Thus, we must enjoy the work of looking at a point in history, culture, objects, and phenomena to identify concepts and put together our own stories. Even though this may not carry any philosophical definition or be meaningful in providing insight to the current era, efforts to continue thinking about even the smallest ideas should go on. Some might say this is

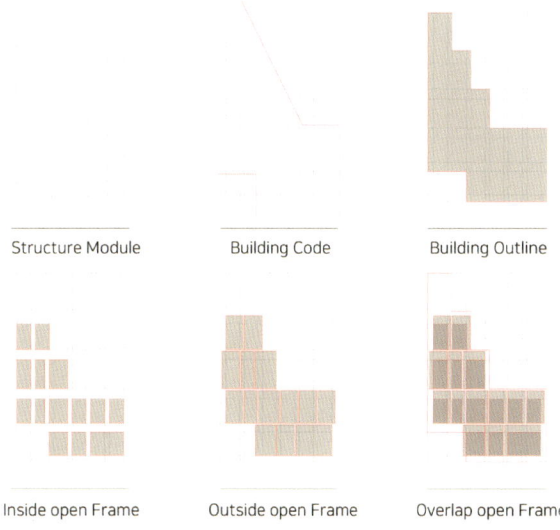

Facade Design

needless from a client's perspective and is meaningless, but only those who are willing to contemplate will find its value.

The buildings on the pedestrian streets of a commercial district often undergo sudden changes along with various trends, such as urban

모어레스 건축 | More Less Architects

오랜 시간 한 자리를 지켜온 상권들이 떠나고 기존 건물은 철거되었다. 주변 거리는 새로운 활기를 띠며 빠르게 물리적으로 변화했다. 상권의 변화 속에서 필요(공간, 기능, 이미지)에 충족하지 못하는 건축물은 쉽게 철거되고 남은 건축물 또한 가로변 파사드 (façade)가 뜯겨져 뼈대(structure)를 내보인다. 이 같은 과격한 행위들은 변화의 시기에 상업가로의 건축물이 맞이하는 아주 자연스런 현상일지도 모른다. 특히 파사드는 상업가로의 특성 속에 건축 내부의 기능적 공간과 별개로 상업가로의 이미지적 요소로서 상징적 역할을 담당하기 때문에 물리적 변화의 주된 대상이 되기도 한다.

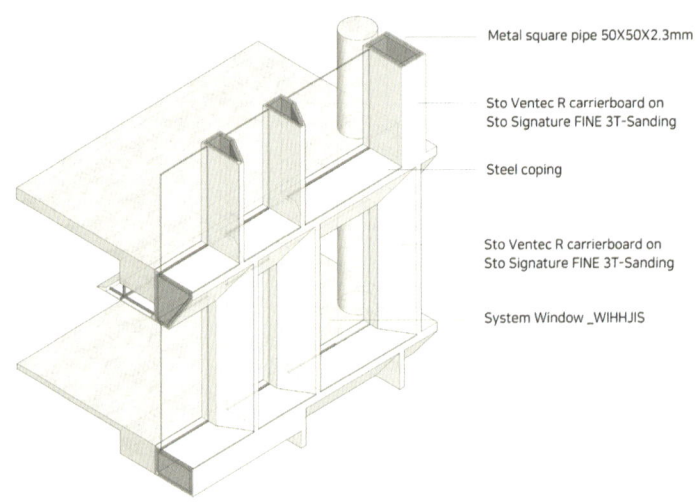

우리가 주목하는 것은 상업가로의 파사드가 공간이라는 본질과 분리될 수 있는 다변적 요소로서 또 다른 가로의 상징이 되어 새로운 기능을 담당하고 있다는 점이다.
파사드는 내부의 기능적 공간과 밀접하게 연계된 형태를 드러내기도 하지만 대부분은 공간과 별개의 독립된 객체로 상업가로에서 역할을 담당한다. 결국 우리의 파사드는 이러한 상업가로 속에서 자유로운 입면(Free façade)으로, 장식된 헛간(decorated shed) 으로 독립된 기능을 만들어가는 것이다.

도산 알로하
Dosan Aloha
ⓒJoel Moritz

structure, society, culture, etc. Depending on the changes in the commercial district, they are required to take on new purposes and new atmospheres, regardless of their intentions at the time of design. The *Aloha Dosan* project in Apgujeong-dong, one of the representative commercial districts, was also carried out in reflection of a trend. Many stores that had been in the same location for a long time left, and existing buildings were demolished. The surrounding streets rapidly changed physically, taking on new vibrant looks. Buildings that do not meet the needs (space, function, image) are easily demolished, and the roadside façades of the remaining buildings are torn down, revealing their structures. Perhaps such rather aggressive actions are a very natural phenomenon faced by buildings on fast-changing commercial streets. In particular, due to the characteristics of a commercial street, a façade often becomes the major target of physical change because it plays a symbolic role as an image element of a commercial street, apart from the functional space inside a building.

Our focus then is that a façade in a commercial street is a fickle element that can be separated from the essence of space, and it can serve as another symbol of the street, taking on a new function. A façade sometimes reveals a form that is closely linked to the functional space of the interior, but in most cases, it plays a role within the commercial street as an independent object separate from the space. In the end, our façade becomes a free façade in this commercial street, creating an independent function as a decorated shed.

More Less Architects

서교동 <티바트 타워(Teyvat Tower)>의 파사드는 기존 건축물을 유지하며 새롭게 부가된 입면이다.

티바트타워
Teyvat Tower
ⓒJoel Moritz

일반 근생의 역할을 담당했던 5층 규모의 건축물은 볼륨이 전면에 돌출되어 다양한 재료와 함께 복잡하게 구성된 형태였다. 그렇게 건축은 앞서 언급한 가로 변화의 흐름 속에 휩쓸려 사옥으로서의 이미지를 위해 입면이 뜯겨진다. 클라이언트의 요구는 특정 기업도 아닌 불특정 사용자를 위한 사옥이다. 상업가로의 파사드는 공간 맞춤형이 아닌 모두를 위한 사옥으로 보편적인 무엇이 되어야 했다. 모든 이들을 위한 넉넉한 기본 티셔츠처럼 말이다. 각 층의 공간을 타이트하게 드러낸 쫄티(slim-fit) 같은 기존 건축물의 표피(skin)를 걷어내고 그사이 비워진 여유(영역)를 부가하며 오버핏(over-fit)의 새로운 표피(skin)를 만든다. 오버핏의 여유로운 공간 속으로 엘리베이터와 발코니를 삽입하고 기존의 복잡한 평면과 입면을 정돈한다. 그러한 오버핏 파사드는 새로운 입면의 질서 속에 건식 공법으로 가볍게 늘어지고 수축하며 공간으로, 발코니로, 또 허공의 가벽으로 드러나게 된다.

Impractical Interests | 무용無用한 관심사

The façade of *Teyvat Tower* in Seogyo-dong is a newly added one that maintains the existing building.

The five-story building that was used for neighborhood facilities was complexly constructed with a mix of materials, with its volume protruding at the front. But, in line with the aforesaid street transformations, the façade of the building is torn down to put on a new image as an office building. The client's request was not for a particular office but for unspecified users. The façade on a commercial street had to be something universal, as an office building for anyone, rather than being space-specific. A good example of this would be a basic T-shirt large enough for everyone. So, we removed the skin of the existing building that was tightly exposing the space on each floor like a slim-fit top and created a new skin of an overfit size by adding empty spaces (areas). Into the overfitted, spacious space, an elevator and balconies have been added, and the existing complex floor plans and elevation are cleaned up. Creating a new order, such an overfit façade is lightly stretched and contracted using a dry construction into a space, a balcony and a free-standing wall.

티바트타워
Teyvat Tower
ⓒJoel Moritz

파사드가 부가된 입면은 <루민연남(Lumen Yeonnam)> 프로젝트로 이어진다. 주거블럭의 가로는 상업거리로 변화하며 상가주택 또한 온전한 상업공간으로서 새로운 파사드의 모습을 기대한다. 상업공간의 투명한 커튼월을 원했던 파사드는 기존 주거 공간 구조를 확장한 개구부의 영역을 넘어 입면 전체에 덧붙여진 형태로 그 모습을 당당히 보여준다. 유리 파사드의 입체적 레이어와 각층에 돌출된 캐노피, 수직 환봉(丸棒) 등은 덧붙여져 장식된 헛간에 질서를 더한다.

루민연남
Lumen Yeonnam
ⓒJoel Moritz

무용無用한 관심사

The elevation with an added façade continues with the *Lumen Yeonnam* project. The streets of residential blocks transform into commercial streets, and even flats-with-shops look forward to having new façades as pure commercial spaces. The façade, which wanted a transparent curtain wall for commercial spaces, boldly displays its new form added to the entire elevation instead of just around the openings expanded structurally from the existing residential space. The multi-level layer of the glass façade, the canopies protruding from each floor, the vertical round bars, etc. all come together to establish order to the decorated shed.

모어레스 건축 | More Less Architects

> 변화하는 상업가로의 입면은 그렇게 기능과 별개의 장식된 헛간으로, 공간보다는 표피에 생각이 담긴다. 형태보다는 질서를 드러내고 그 속에 두께를 더하고 질감을 더하려고 한다. 표피의 실용성을 넘어 무용한 건축의 감각을 탐구해간다.

루민연남
Lumen Yeonnam
ⓒJoel Moritz

Impractical Interests | 무용무용한 관심사

The façade of a commercial street in transformation is like a decorated shed, apart from its functionality, with ideas embedded in its skin rather than the space itself. More than the form, order is revealed, with thickness and texture added to it. Going beyond the practicality of the skin, an impractical sense of architecture is explored.

More Less Architects

부분에서 전체

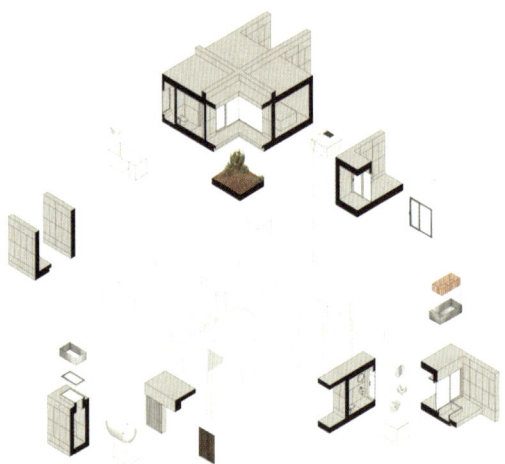

우리는 때론 건축의 작은 것에, 아주 부분적이고 사소한 것에 몰입하려고 애쓴다. 작고 부분적인 것에 건축의 모든 것을 담으려는 듯 말이다. 잘 인식되지 않는 것에 집착을 보이고 선량한 현장에 불안을 조성하며 현장소장을 붙잡고 민망하게 열을 내기도 한다.

건축가는 매우 다양한 건축 요소를 다룬다. 중요한 기능을 담당하는 요소가 있는 반면 별 기능 없이 잘 인식되지 않는 요소도 많다. 그럼에도 그 사소한 요소들이 모여 건축의 가치를 만든다. 주요한 기능은 실용적이고 합리적인 역할을 담당하지만 사소한 마감의 처리는 무용한 분위기를 담아내기에 좋다. 그렇게 재료가 만나는 곳에, 시선이 흐르는 곳에, 손끝이 닿는 곳에 우리는 많은 에너지를 쓴다. 건축가가 의미를 부여한 사소한 것이 고객과 시공자에게 공유될 때 건축의 모든 것은 완성된다.

가온누리
Gaon-Ruri
ⓒJoel Moritz

Impractical Interests

From Parts to Whole

Sometimes we sweat the small stuff, the things that are very partial and insignificant parts of architecture, as if to contain everything about architecture in the small and partial. We obsess over things that are not well-recognized, create anxiety among workers, and even embarrassingly blow up at the site manager over nothing.

Architects deal with a wide range of architectural elements. Among the elements, some have important functions while many others have no particular function and are easily overseen. Nevertheless, those minor elements come together to create the value of architecture. While the main functions play a practical and rational role, it's the minor finishing touches that capture the atmosphere of futility. We spend a lot of energy in areas where materials meet, which our eyes want to see and our fingertips touch. When the little things in which the architect has inspired meaning are shared with the client and constructor, all is complete in architecture.

전라남도 광양 <가온누리(Gaon Ruri)> 재가노인복지시설은 구도심과 공장 부지 사이의 아름다운 자연환경을 마주하고 있다. 공장지대 앞으로 수어천이 흐르고 그 너머 펼쳐지는 풍경은 광양 진월의 천왕산 절경이다. 우리는 자연의 아름다운 풍경이 계절마다 변화하며 공간을 가득 채우길 바랐다. 어르신들이 사용하는 공간이기에 무엇보다 풍경이 중요했고 건축의 모든 요소들도 그것으로부터 출발했다. 천왕산 풍경을 파노라마로 담아내기 위해 가능한 기능 공간을 통합하여 배치하고, 구조는 풍경을 향한 시선에 방해되지 않도록 최소한의 원형 기둥(ø114.6mm)으로 가볍게 해결했다. 창호 프레임(54mm)도 기둥 열 뒤로 숨기고, 핸드레일은 기둥에 고정하여 부가적인 요소를 줄였다. 가능한 창의 인방을 줄이고 몰딩은 없앴으며 가벼운 실링만 파노라마 풍경을 위해 간결한 수평선을 만들게 했다. 테라스 앞 담장도 도시의 불편한 시선을 차단할 만한 적절한 높이로 제안했다.

한정된 예산으로 시작한 프로젝트지만 클라이언트, 시공사, 협력사 모두가 우리의 가치를 공유하며 그 열정에 공감했다. 꽤 먼 현장을 수차례 비행기와 렌트카로 오갔고, 플랜트 시공이 전문인 지역 시공사는 까다롭기만한 우리 요구를 하나하나 차근차근 풀어 나갔다. 아주 사소한 영역까지, 우리의 예민함까지 온전히 공감하지 못하더라도 모두가 끝까지 노력하고 인내해주었다. 그렇게 건축이 되었다. 우리 모두가 <가온누리> 재가 노인복지시설을 찾은 어르신들이 공간 속에서, 풍경 속에서 평온함을 느끼길 바랐다.

가온누리
Gaon-Ruri
ⓒJoel Moritz

제주 비자림에 가면 붉은 화산송이가 펼쳐져 있다. 자연이 만들어낸 붉은 빛깔이 좋았다. 제주는 강인한 자연의 섬이다. 제주를 돌아보면 어떤 형태든 자연만 못하고 어떤 재료든 자연 앞에서 나약하다. 건축은 단순해야 한다. <수리움(Surium)>은 붉은 화산송이의 땅 위에 붉고 단단한 덩어리가 놓인 형상이다. 건축은 제주의 푸른 하늘과 초록의 자연 사이에서 하나의 오브제 같은 덩어리이길 바랐다.

Impractical Interests | 무용無用한 관심사

Gaon Ruri Home Care Service Center for Seniors in Gwangyang, Jeollanam-do is situated in a beautiful natural environment between the heart of an old city and a factory site. Sueocheon (tributary) flows in front of the factory, and the scenery beyond the tributary is a spectacular view of Mt. Cheonwang in Jinwol, Gwangyang. We wanted the splendid scenery of nature to fill the space as it changes by season. Since the end-users of the space would be elderly members of the community, the scenery was more important than anything else and thus, all elements of the architecture started from there. Functional spaces were integrated and arranged to capture the panoramic view of Mt. Cheonwang, and for the structure, we kept round pillars (ø 114.6 mm) to a minimum so as not to interfere with the view toward the landscape. The vertical frames (54 mm) of the windows were then hidden behind the row of pillars, and the handrails were fixed to the pillars to reduce additional elements. We also limited the number of window lintels and did not use moldings, and created a simple horizontal line for a panoramic view only with the light ceiling. In addition, we proposed to keep the height of the fence in front of the terrace to be just tall enough to block uncomfortable views of the city.

Although this was a project with a limited budget, our values were supported by the client, contractor, and partners, and we were all on the same page. It took several visits to the fairly distant site by plane or rental car, and the local contractor specializing in plant construction gradually worked out each and every one of our complicated requests. Even though they couldn't empathize with us down to the smallest details and even our sensitivities, everyone worked hard and stayed patient to the end. That's how it was built. With a single hope that the seniors at *Gaon Ruri* would feel peaceful in the space and in the scenery.

There are red volcanic clusters at Bijarim Forest in Jeju. I was attracted to the red color created by nature, which is inherently strong in Jeju. Looking around Jeju, nothing surpasses nature and all becomes powerless in the face of nature. Architecture should be simple. Surium takes the form of a red, hard cluster placed on a red volcanic ground. I wanted *Surium* to be seen as an object-like mass in between Jeju's blue sky and green nature.

수리움
Surium
ⓒJoel Moritz

제주의 화산 송이석처럼 붉은 덩어리를 만들기 위해 우리는 컬러콘크리트를 감당할 레미콘 공장을 찾아야 했고 두겁대 없이 방수를 해결해야 했다. 또 덩어리는 바닥의 붉은 화산 송이석과 만나야 했기 때문에 기단부를 일일이 쪼아내어 거친 질감을 드러냈다. 제주 로컬 시공사는 열정을 갖고 참여해 주었다. 그들에게 익숙하지 않은 마감들이기에 모든 분야별 작업자와 협의가 필요했다. 공사가 마무리되기까지 현장소장이 세 번 바뀌었지만 그럼에도 남은 이들은 끝까지 포기하지 않고 에너지를 쏟아주었다.

수리움
Surium
ⓒJoel Moritz

In order to resemble the red mass like Jeju's volcanic rock, we needed to look for a ready-mix concrete factory that could make colored concrete and keep water out without using parapets. Because the mass had to be in contact with the red volcanic pumice rocks on the ground, the base foundation was individually chiseled away to reveal its rough texture. The local contractor from Jeju enthusiastically participated in the project. Because the finishing methods were unfamiliar to them, they needed to consult with the workers of each trade. By the completion phase, the site manager had changed three times, but those who hung on gave their all into the project.

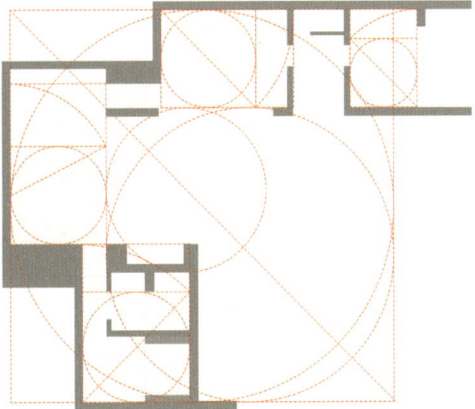

수리움 Surium ⓒJoel Moritz

More Less Architects

<수리움>의 개념적인 덩어리는 각기 다른 성격과 크기의 공간으로 구성되어 적절히 영역을 가지며 연결된다. 외부공간도 하나의 분위기를 만들며 대지의 효과를 최대화한다. 단기 체류를 위한 스테이는 외형이 닫힌 덩어리 같지만 내부는 다양한 건축적 단면을 갖는다. 바닥 레벨 차이를 이용해 시선을 낮춘 거실, 소리의 울림을 위한 높은 볼트 형태의 다이닝, 꺾인 천창을 통해 빛을 담아내는 주방, 좁고 낮은 공간의 침실, 외부 수영장에서 이어지는 내부 욕조, 기타 다양한 건축 공간 등은 일상의 공간적 경험을 더욱 풍요롭게 한다. 우리는 이곳을 찾는 사람들이 단면적 건축을 경험하며 분위기 있는 공간을 통해 무뎌진 공간 감각을 회복하길 바랐다.

수리움
Surium
ⓒJoel Moritz

건축을 구성하는 요소들을 하나하나 관심 갖고 들여다보면 우리가 다룰 수 있는 많은 것들이 있다. 새로운 재료와 구축법도 다양하지만 일상적인 재료와 구축법 속에도 우리가 고민할 거리는 많다.

브리크둔촌
Brick Dunchon
ⓒJoel Moritz

Impractical Interests

The conceptual mass of *Surium* is composed of spaces that are different in characteristics and scales, each appropriately zoned and connected. The external space has its own atmosphere, too, maximizing the effect of the site. While the exterior of this building for short-term stays appears to be one big chunk, the interior contains various architectural cross-sections. The sunken living room, the dining area with a high barrel vault ceiling for echoing sounds, the kitchen that captures light through a curved skylight, the bedroom in a narrow and low space, the indoor bathtub that leads to the outdoor swimming pool, and other various architectural spaces further enrich the spatial experiences of everyday life. Our hope was that the visitors

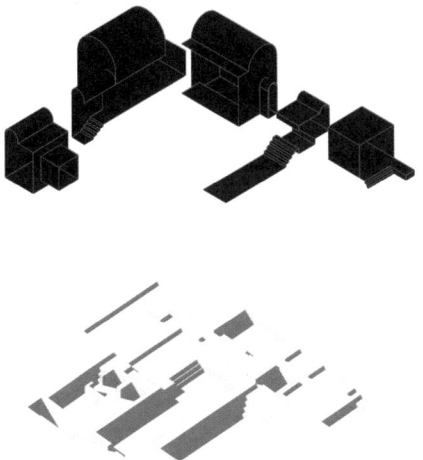

to this place would experience the cross-sectional architecture and recover their dull spatial awareness.

Taking a careful look at each of the elements that make up architecture, we come to learn that there are many things we can work with. While there are new materials and numerous construction methods, we can find plenty of ideas to ponder upon using everyday materials and common construction methods.

<브리크둔촌(Brick Dunchon)>에서는 주거건축이 가진 실용적이고 기능적 공간계획 외에도 외벽 마감재인 벽돌의 쓰임에 우리의 고민을 담아내려 했다. 오랜 시간 주거 마감재로 쓰인 벽돌은 그 자체로 분위기가 있다. 하지만 우리는 벽돌의 새로운 감각을 드러내고자 했다. 벽돌은 구조적 역할에서 치장의 역할로 넘어온 지 오래다. 우리는 벽돌이라는 재료의 물성과 구축 방식을 통해 그것의 표피가 만들어낼 수 있는 분위기를 상상했고 색, 크기(온장과 반장), 줄눈 처리방식, 깨진 단면 등 가능한 요소의 조합을 통해 상상한 건축적 분위기를 만들어갔다.

브리크둔촌
Brick Dunchon
ⓒJoel Moritz

상층부의 어두운 반파 벽돌은 깨진 단면의 거친 질감과 깊은 줄눈을 통해 많은 그림자를 드리운다. 하부로 내려오면서 온장과 반파 벽돌이 결합하며 표면의 질감은 부드러워지고 그림자도 줄어든다. 하부의 밝은 온장 벽돌은 매끄러운 표면을 드러내고 줄눈은 표면까지 메우는 방식으로 그림자를 사라지게 한다. 치밀하게 다뤄진 벽돌의 표피는 색상과 단면의 질감, 빛의 그림자로써 의도한 분위기를 자아낸다.

Impractical Interests | 무용無用한 관심사

As for *Brick Dunchon*, we tried to reflect our concerns in using bricks, an exterior building material, along with the practical and functional spatial planning for residential architecture. Bricks, which have been around for a long time as a building material for homes, have a unique atmosphere of its own. But we wanted to try something else with bricks. Bricks have long since moved from a structural role to a decorative role. Identifying the material properties of bricks and the construction method, we imagined the atmosphere that the skin of the bricks can bring out, and by combining possible elements, such as colors, sizes (bricks and half-brick), jointing methods, broken brick faces, etc., we were able to achieve the imagined architectural atmosphere.

The dark, half-brick on the upper part of the building cast many shadows through the rough texture of the broken sections and recessed joint. The surface on the lower part of the building, on the other hand, is softer with less shadows as both bricks and half-brick are used. Lighter-colored bricks on the lower part have a smooth surface, and by filling the joint flush with the brick surfaces, shadows disappear. The carefully handled skin of the bricks creates the intended atmosphere by color, cross-sectional texture, and light shadows.

브리크둔촌
Brick Dunchon
ⓒJoel Moritz

몇 가지 프로젝트를 통해 언급한 바와 같이 우리가 심도 있게 다루는 대부분의 것들은 건축의 부분적인 것에 불과하며 그것조차도 실용적인 현대 사회에 유용하게 논의되는 것이 아니라는 것을 잘 안다. 우리의 생각은 거창해 보이지만 사소하고, 유용할 것 같지만 무용하다. 우리가 구축한 공간은 상상과 경험을 통해 획득한 분위기로서 이것은 개인적 감각에 한계가 있다. 하지만 우리는 뻔한 방식에서 벗어나 공간과 재료를 자유로이 탐구하고 새로이 구축해감으로써 소소한 디테일을 넘어 건축의 또 다른 가치를 만들고 싶다.

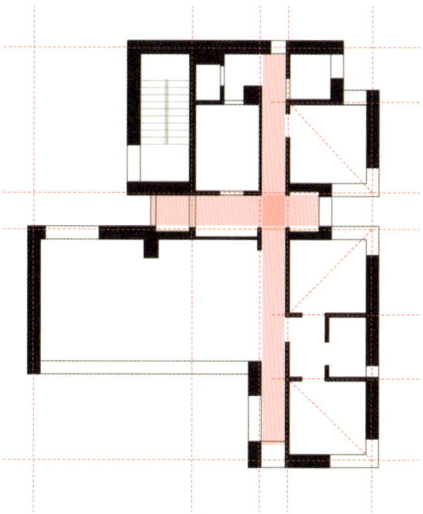

As pointed out in some of our projects, most of the things we pay attention to are only a part of architecture, something not even productively discussed in a pragmatic modern society. While our thoughts seem grand, they are trivial, and they seem practical but impractical. The space we build is an atmosphere acquired through imagination and experience, and thus it has limitations in terms of our personal senses. But we want to go beyond what's obvious and freely explore spaces and materials, discovering new ones and creating another value in architecture beyond just minor details.

상상과 확신

바둑 고수였던 한 친구가 내게 바둑의 몇 수 앞까지 상상할 수 있는가를 물은 적이 있다. 이러한 상상은 가만히 머릿속으로 가상의 수를 두어 가며 상대의 수를 계산하는 방법으로, 사실 그를 한번이라도 이기기 위해 내게 필요한 것은 상상력이었다. 하지만 상상은 귀찮고 피곤한 일이기에 안타깝게도 나의 뇌는 대부분 뒤틀리고 왜곡되어 현실과 다른 적당한 답을 준다.

옛 지식인들은 입체 도형을 상상하는 연습을 반복하며 어느 순간 복잡한 기하학적 도형을 머릿속으로 충분히 구현해내는 능력을 가질 수 있었다고 한다. 입체 도형을 머릿속으로 360도 돌려보고 다시 그것을 전개도로 펼쳐 보이는 것이다. 도대체 상상하는 연습을 얼마나 해야만 했을까?

건축가는 건축 공간을 수없이 상상해 간다. 계획하면서 공간의 스케일과 분위기를 상상하고 공간의 흐름과 형태의 비율을 상상한다. 더 나아가 구조와 접합부를 상상하고 마감과 마감의 조합을 상상한다. 분명 쉽지 않은 일이다. 최근 틀어진 기둥을 계획하며 구조와 마감이 머릿속에 명확히 그려지지 않아 난감한 적이 있었다. 우리의 상상과 현실의 싱크로율(synchronization)은 과연 어떠한가? 상상의 수준이 계획의 확신과 결과의 수준을 높인다는 사실은 분명하다.

건축을 계획하다 보면 다양한 단계에서 스스로 확신을 가져야 하는 순간들이 있다. 건축은 정답이 없기에 확신이 필요하다. 의도했던 많은 건축 요소가 각각의 의도대로 작동하고 모든 것이 의도한 공간감으로 드러날 것이라는 확신, 그것을 위해 계획 과정에서 수많은 상상의 과정과 검증의 과정을 거친다. 건축가는 누구나 각자 확신의 단계와 방법이 있을 것이다.

우리는 계획 초기 여러 요구사항과 환경적 조건, 아이디어를 마음껏 펼쳐 고뇌의 시간을 보낸다. 오랜 시간 끝에 공간 구조와 디자인이 결정되면 잠시 휴식과 검증의 시간이 필요하다. 긴 시간의 열정은 때론 편협한 생각과 독단적 판단을 만들고, 빠듯한 시간은 본능적 감각만을 남기기도 한다. 계획이 길어질수록 지난 시간의 미련과 변화의 아쉬움도 있다. 모형을 만들 시간이다. 긴 시간 빠져 있던 생각의 고리를 끊고 깨끗하게 머릿속을 비워낸다. 모형을 만드는 시간은 하나의 휴식과 같다. 계획에 따라 재료를 자르고 붙이다 보면 어느 순간 생각은 사라지고 단순한 몸의 움직임만 남는다. 그러한 시간을 보내며 나의 무의식은 객관적으로 작업을 바라보게 된다. 그렇게 모형이 서서히 완성되어갈 때쯤 스스로 확신이 생겨난다. 그래서 나는 직접 모형을 만든다.

"이것으로 되었다!" 혹은 "이건 아니다!"

Imagination and Confidence

A friend who was a Go expert once asked me how many moves I could possibly predict in Go. This required making imaginary moves in my head and calculating my opponent's moves, thus imagination was all I needed to win even once. But sadly, my brain mostly gives me a cursory answer distorting reality, because imagining is cumbersome and tiring.

They say our ancient scholars were able to fully embody complex geometric figures in their minds after repeated practices of imagining three-dimensional figures. It's like rotating a three-dimensional figure 360 degrees in your head and unfolding it again as a planar figure. How much did they possibly practice imagining?

Architects imagine architectural spaces countlessly. While planning, we imagine the scale and atmosphere of a space, along with the flow of space and the proportions of forms. Furthermore, we imagine the structures, joints, finishes and the combination of finishes. It's definitely not easy. Recently, I was working on a twisted pillar and had trouble coming up with the appropriate structure and finishes. To what extent does our imagination synchronize with reality? It's true that the level of imagination increases the level of confidence in planning and results.

When planning, there are moments when you need to be confident in yourself. Because there is no one correct answer in architecture, confidence is what you need: confidence that many architectural elements will work as intended and that the space will be revealed as intended. For this, we go through numerous processes of imagination and verification. Every architect will have their own steps and methods of assurance.

The early stages of planning are often spent agonizing over various requirements, environmental conditions, and ideas. After hours of thinking, the structure and design are decided, and that's when we need to take a break and validate the proposal. Long hours of enthusiasm sometimes lead to narrow-mindedness and dogmatic judgment, and a tight schedule sometimes leaves only instinctive senses. The longer the planning, the more you regret and the more changes you want to make. It's time to make a model. It's time to break the chain of thoughts that have been in your head for a long time and clear your mind. The time you spend making a model is like taking a break. As you cut and paste materials as planned, wandering thoughts go away and the only thing that is at work is your body. It is through those times that my unconscious takes an objective look at my work. Then toward the completion of the model, I am convinced. This is why I make models by myself.

"This is it!" or "This isn't it!"

설마 시간이 남아서 모형을 만들겠는가? 그대들은 알고 있으리라. 상상한 부분을 가능한 원형 그대로 만들어 빛이 잘 드는 창가에 올려놓고 감상하고 싶었던 것을. 그 순간 모형이 현실 속에서 실현되는가는 중요한 문제가 아니다. 우리는 모형을 바라보며 흐뭇하게 충만한 기분을 만끽하고 싶었던 것이다. 그러한 감상의 끝에, 우리는 우리의 설계에 확신을 갖는다.

무용無用한 관심사

Do I make models because I have time to kill? Architects will know the answer. It would be to make the model imagined in your head as close to the original as possible and enjoy it on a windowsill under the sunlight. At the moment, it doesn't matter whether the model will be realized in reality. It's the satisfaction and fulfillment coming from gazing in contentment. It is such reflection that helps us feel confident in our design.

도락(道樂)의 마음

현시대의 건축 공간은 추운 풍류를 논하기도, 불편한 낭만 따위를 찾기도 힘들다. 보편적 건축으로서 대표적인 주거(아파트)는 금 덩어리와 같이 시세를 쫓는 상업적 수단이 되었고 그 속에서 공간은 인큐베이터와 같이 안락한 면적(m^2)*만 남겨 놓았다. 이 같은 공간은 우리에게 보편적인 삶의 기준이 되었다. 공간의 주요한 가치가 면적(m^2)으로, 방의 개수로 쉽게 정의되는 것이 이 시대의 주거 건축일지도 모른다.
[면적(m^2): 공간의 이차원적인 해석]

서양의 기술과 사고 아래 물리적 공급으로 시작된 획일화된 주거공간은 우리에게 '편리함'의 상징적 공간이 되어버렸다. '편리함'의 경험은 빠르게 공간에 대한 '익숙함'으로 변해 갔고 그러한 주거는 우리에게 '보편화된 건축'이 됐다. 이제 우리는 일상의 삶의 의미로서 '보편화된 건축'의 가치를 논하기보다는 기능에 충실한 익숙한 공간으로 고민 없이 받아들이며 쉽게 적응하고 학습해 가는 듯하다. '보편화된 건축'은 편리할 수 있겠으나 좋은 공간이라 말할 수 있을지는 의문이다. 지난 시대의 서양적 사고의 공간에 어느덧 현 시대 우리의 일상을 끼워 맞추며 살아가는 것은 아닐까 고민해 보아야 한다.

건축공간은 시대의 흐름에 따라 변화와 발전이 없는 것도 아니며 현대 주거의 안락함, 편리함을 부정하거나 그 기능과 역할을 비하하자는 것도 아니다. 다만 현시대 건축의 방향은 점점 더 면적(m^2)과 실용성(에너지, 기능)이라는 유용한 대의를 위한 가치를 논할 뿐, 공간 속 무용한 가치에 대해 그리 의미를 두지 않는다는 것을 환기하고 싶다. 건축가들의 감각적 작업은 있겠으나 그렇다고 철저히 고민되거나 설명되는 것도 아닌 듯하다.

어린 시절 툇마루에 앉아 처마끝에서 떨어지는 빗방울을 바라보는 것이 좋았다. 내게 좁은 마당은 계절을 담는 설레는 공간이었고, 한지 창으로 묘한 빛을 채우던 방은 평온함을 선사하는 공간이었다. 육중한 문을 힘겹게 열고 들어간 어느 교회의 어둠 속 공간과 천창의 빛 그리고 빛이 자아내는 질감은 나를 숭고함으로 이끌었다.

쌔미후암
SSamy Huam
ⓒJoel Moritz

Heart of Amusement

In today's architectural space, it is not easy to discuss cold elegantly or find discomfort romantically. The most common form of universal architecture, housing (apartments) has become a commercial means to chase the market price like a gold nugget, leaving space to become nothing more than just an area (sq. m).* Spaces like these have become a universal standard of living for us. Simply defining the main value of space by area measurement (sq. m) and the number of rooms may be the residential architecture of this era [Area (sq. m): Two-dimensional interpretation of space].

The standardized residential space that began as a physical supply under Western technology and thinking has become a symbolic space of "convenience" for us. The experiences of "convenience" has quickly evolved into "familiarity" of space, and such housing became "universal architecture" for us. Now, it seems like we are easily adapting and learning, accepting it without hesitation as a familiar space faithful to its function, rather than discussing the value of "universal architecture" as a significance in our everyday life. While such an architecture may be convenient, we need to ask ourselves whether it can be considered a good space. We must think about whether we are living a lifestyle that has been fitted into the space of Western thinking of the past era.

Architectural space does not mean no change or development over time; this also does not mean that we deny the comfort and convenience of modern housing or denigrate its functions and roles. I do, however, want to point out that the direction of present-day architecture is increasingly focusing on useful purposes, such as areas (sq. m) and practicality (energy and functions), and is not placing much emphasis on what is considered impractical within space. Some works by architects may be sensuous, but they do not seem to have been thoroughly thought out or explained.

When I was young, I enjoyed sitting on toenmaru (Korean traditional veranda) and watching the drops of rain falling from the edges of eaves. To me, the small yard was an exciting space where I could experience all four seasons, and the light that transmitted through Hanji (traditional Korean paper) windows made the room ever so cozy and tranquil. The dark space of a church with an oversized door I barely managed to open led me to a feeling of sublimity under the light from the skylight and the texture created by the light.

꽉 채운 용적률이나 135mm 가등급 단열재의 위력도 중요하지만 때로는 일상 속 빛이, 한날의 빗소리가, 어느 공간의 울림이 삶에 더 중요할 수 있다. 현대의 가치 기준에서는 불편한 처마의 빗방울이나 무거운 문, 추운 툇마루 같은 건축 속 풍류들이 실용성의 판단 아래 대부분 사라진다. 처음부터 의미 없는 요소일 수도 있지만 누군가에게는 순간순간 느껴지는 공간의 기억이기도 하다.

    건축은 유용한 것과 무용한 것의 조화에 있을 것이다. 아직 우리 시대의 건축은 대부분 이론이나 유용한 것에 치중되어 있다. 분명 건축의 규제는 유용한 것에 기준을 둘 수밖에 없다고 하나 무용한 가치에 대한 논의는 다른 형식으로 설명되고 공유되어야 하지 않을까. 우리는 그것을 찾고 설명하는 데 소홀하고 나태하지는 않은가?

    르 코르뷔지에도 <롱샹>과 <작은 집(어머님의 집)>에서는 현대 건축의 유용한 이론보다 무용한 것에 더 집중했을 것이다. 근대건축의 5원칙만큼 <작은 집>에 설계된 호수를 바라보는 고양이를 위한 작은 테라스가 내겐 아름다움이다.

    유용한 것에 치중한 단편적인 해석과 규제는 공간 다양성의 부재를 만들었고 그 속에서 살아가는 우리의 일상을 단순하게 이끌어 온 것 또한 사실이다. 그렇게 지속되어온 보편적 공간은 '편리함'이라는 잣대가 되어 새로운 다양성의 공간을 '불편함'이라는 영역에 놓아 두게 한다. 불편함에 대한 인식은 우리에게 공간적 경험의 한계를 만들고 점점 더 무용한 많은 가치들을 사라지게 한다.

    좋은 공간 속에는 신체적 편리함과 함께 생각과 감각을 이끌 수 있는 불편한 공간이 필요하지 않을까? 편리함만 추구하는 현 시대에 더욱더 고민해야 할 부분은 아닐까? 누군가는 실험이라 말할 수 있고 또 그것이 폭력이라 말할 수 있겠으나, 현시대의 기능적이고 획일화된 공간만한 폭력이 또 있을까? 추운 툇마루는 때론 불편함이 될지언정 우리에게 삶을 풍요롭게 하는 요소임은 분명하다. 그 무용한 가치를 짧은 말로 논하기는 어려우나 작은 경험에서도 우리에게 깊이 배어있는 감성은 드러나리라 생각한다. 경험의 부재일 뿐 우리 감각의 본질은 그리 단순하지 않다.

현 시대성에 대해 건축 공간으로 고민해야 할 많은 것들이 있다. 빠른 변화에 맞춘 기술도, 라이프스타일의 변화도, 새로운 프로그램의 수용도 중요하지만 한편으로는 우리에게 깃든 감각의 무용한 가치를 돌아보고 새로운 공간으로 보여주는 시도가 필요하리라. 나는 이것이 다양성으로 표출되길 바란다.

    물론 이러한 다양성은 단지 좋은 건축을 말하는 것도 아니고 또 이것을 강요할 문제도 아니다. 하지만 새로운 공간의 시도가 불편한 경험을 요구하고 공간의 새로운 가치를 불러 일으키게 할 것이라 데는 의심의 여지가 없다.

Of course, the floor area ratio and the best grade insulation of 135 mm are all important, but just sometimes, the light that brightens an ordinary day, the sound of rain, or the echoes that fill a space could be more important in life. As per the modern standards of value, most architectural elements that are considered inconvenient, under the judgment of practicality, disappear, such as the raindrops on the edges of eaves, overweight doors, and cold toenmaru. While they may be meaningless from the beginning, for some, they may bring back memories of space moment by moment.

Architecture may be a harmony of practical and impractical things. But today's architecture still leans more toward theories or practicality. It is understandable that architectural regulations have no choice but to be based on what is useful but shouldn't impracticality be discussed and shared as well? Aren't we negligent and indolent in finding and explaining it?

In *"the Ronchamp chapel"* and *"A Little House (Mother's House),"* Le Corbusier may also have focused more on the impractical aspects of modern architecture rather than on practical theories. As much as the five points of modern architecture, the small suspended platform that allowed the cat to enjoy the view of the lake in Le Corbusier's *"A Little House"* is a beauty to me.

The reality is that fragmentary interpretations and regulations only focusing on practical things have created the absence of spatial diversity and have also simplified our daily lives within it. The universal space that has continued in this way has become a standard of "convenience" and places new spaces of diversity in the realm of "inconvenience." Our perspective on inconvenience creates limitations in our spatial experiences and increasingly causes many impractical values to disappear.

Shouldn't a good space possess both physical convenience and an uncomfortable space that can guide thoughts and senses? Isn't this something we need to think more about in this era in which we only pursue convenience? For some, this may be an experiment or a form of violence, but what could be more violent than the functional and standardized space of the present era? Although the cold toenmaru may be inconvenient at times, it is certainly an element that enriches our lives. It wouldn't be right for me to discuss its impractical value in a few words, and I believe that even small experiences will reveal the emotions that are deeply ingrained in us. The nature of our senses is not so simple; it is just an absence of experience.

There are many things to consider about the nature of this era in terms of architectural spaces. Technology adapted to rapid changes, lifestyle changes, and acceptance of new programs are all important, but on the other hand, there is a need to look back on the impractical values of our senses and make efforts to reflect them in a new space. I hope this is expressed in diversity.

Of course, this diversity is neither just limited to good architecture nor is it a matter of forcing it. There is, however, no doubt that such attempts will require uncomfortable experiences yet will bring about new values in space.

프로젝트

수리움
- 위치: 제주특별자치도 제주시 한경면 청수리
- 용도: 단독주택
- 책임건축가: 김영수
- 디자인팀: 송채윤
- 대지면적: 475㎡
- 건축면적: 116.49㎡
- 연면적: 116.49㎡
- 건폐율: 24.52%
- 용적률: 24.52%
- 규모: 지상 1층
- 구조: 철근콘크리트
- 외부마감: 컬러콘크리트
- 내부마감: 테라코트, 타일
- 준공연도: 2022
- 구조설계: 혜담구조
- 기계설계: 성지 E&C
- 전기/통신설계: 아이에코 ENG
- 조경설계: 그린샐러드 플라워&가든
- 시공: 유진건설
- 사진: 조엘 모리츠

나지요네
- 위치: 제주특별자치도 제주시 한경면 청수리
- 용도: 단독주택
- 책임건축가: 김영수
- 대지면적: 871㎡
- 건축면적: 84.52㎡
- 연면적: 84.52㎡
- 건폐율: 9.7%
- 용적률: 9.7%
- 규모: 지상 1층
- 구조: 철근콘크리트
- 외부마감: 노출콘크리트
- 내부마감: 콘크리트 폴리싱, 미장 위 도장 마감
- 구조설계: 이든구조
- 기계설계: 성지 E&C
- 전기/통신설계: 아이에코 ENG
- 시공: KR건설
- 사진: 이택수

도산 알로하
- 위치: 서울시 강남구 신사동 654-8
- 용도: 근린생활시설
- 책임건축가: 김영수
- 대지면적: 273.7㎡
- 건축면적: 161.0㎡
- 연면적: 730.48㎡
- 건폐율: 58.83%
- 용적률: 199.30%
- 규모: 지하 1층, 지상 5층
- 구조: 철근콘크리트
- 외부마감: 스토 벤텍 R, 스토 시그니처, 석재
- 내부마감: 노출콘크리트 위 면 처리
- 준공연도: 2021
- 구조설계: 이든구조
- 기계설계: 이레 E&C
- 전기/통신설계: 아이에코 ENG
- 시공: 지음씨엠
- 사진: 조엘 모리츠

브리크둔촌
- 위치: 서울시 강동구 둔촌동 107-3
- 용도: 단독주택
- 책임건축가: 김영수
- 대지면적: 293㎡
- 건축면적: 139.26㎡
- 연면적: 296.89㎡
- 건폐율: 47.53%
- 용적률: 101.33%
- 규모: 지상 3층
- 구조: 철근콘크리트
- 외부마감: 벽돌
- 내부마감: 석고보드 위 도장
- 구조설계: 혜담구조
- 기계설계: 성지 E&C
- 전기/통신설계: 아이에코 ENG
- 시공: 지음씨엠
- 사진: 조엘 모리츠

루민연남
- 위치: 서울시 마포구 연남동 567-15
- 용도: 근린생활시설
- 책임건축가: 김영수
- 디자인팀: 양원중
- 대지면적: 194.4㎡
- 건축면적: 116.25㎡
- 연면적: 468.64㎡
- 건폐율: 59.8%
- 용적률: 188.9%
- 규모: 지하 1층, 지상 4층
- 구조: 일반철골구조 및 철근콘크리트
- 외부마감: 유리, 금속, 스타코
- 내부마감: 석재타일, 미장 위 도장
- 구조설계: 혜담구조
- 시공: 지아택건설
- 사진: 조엘 모리츠

사라한남
- 위치: 서울시 용산구 한남동 792-2
- 용도: 근린생활시설
- 책임건축가: 김영수
- 대지면적: 123.4㎡
- 건축면적: 73.10㎡
- 연면적: 310.12㎡
- 건폐율: 59.23%
- 용적률: 198.36%
- 규모: 지하 1층, 지상 5층
- 구조: 철근콘크리트
- 외부마감: 커튼월, 아연도 철판, 노출콘크리트
- 내부마감: 아연도 철판, 노출콘크리트
- 준공연도: 2021
- 구조설계: 이든구조
- 기계설계: 성지 E&C
- 전기/통신설계: 아이에코 ENG
- 시공: 센트럴건설
- 사진: 조엘 모리츠, 최진보

쌔미후암
- 위치: 서울시 용산구 후암동 58-19
- 용도: 상가주택
- 책임건축가: 김영수
- 디자인팀: 송채윤
- 대지면적: 111.4㎡
- 건축면적: 63.8㎡
- 연면적: 180.8㎡
- 건폐율: 57.27%
- 용적률: 162.298%
- 규모: 지상 4층
- 구조: 철근콘크리트
- 외부마감: 벽돌
- 내부마감: 석고보드 위 도장, 노출콘크리트
- 준공연도: 2022
- 구조설계: 혜담구조
- 기계설계: 성지 E&C
- 전기/통신설계: 아이에코 ENG
- 시공: 솔하임건설
- 사진: 조엘 모리츠

티바트빌딩
- 위치: 서울시 마포구 서교동 325-25
- 용도: 근린생활시설
- 책임건축가: 김영수
- 대지면적: 171.9㎡
- 건축면적: 87.53㎡
- 연면적: 487.92㎡
- 건폐율: 50.92%
- 용적률: 228.64%
- 규모: 지하 1층, 지상 5층
- 구조: 철근콘크리트
- 외부마감: 스타코
- 내부마감: 석고보드 위 도장, 아연도 철판
- 준공연도: 2023
- 구조설계: 혜담구조
- 시공: 지음씨엠
- 사진: 조엘 모리츠

Projects

SURIUM
- Location: Cheongsu-ri, Hangyeong-myeon, Jeju-si, Republic of Korea
- Program: Single family house
- Architects: Youngsoo Kim
- Design team: Cheayoon Song
- Site area: 475m²
- Building area: 116.49m²
- Gross floor area: 116.49m²
- Building coverage ratio: 24.52%
- Floor area ratio: 24.52%
- Number of levels: 1F
- Structure: Reinforced concrete
- Exterior Finish: Color concrete
- Interior Finish: Terraco, Tile
- Year completed: 2022
- Structural engineering: Hyedam Structure
- Mechanical engineering: SeongJi E&C
- Electrical engineering: I-eco ENG
- Landscape desing: GreenSalad Flower&Gardening
- Construction: Yujin Construction
- Photography: Joel Moritz

NASILLONNER
- Location: Cheongsu-ri, Hangyeong-myeon, Jeju-si, Republic of Korea
- Program: Single family house
- Architects: Youngsoo Kim
- Site area: 871m²
- Building area: 84.52 m²
- Gross floor area: 84.52 m²
- Building coverage ratio: 9.7%
- Floor area ratio: 9.7%
- Number of Levels: 1F
- Structure: Reinforced concrete
- Exterior Finish: Exposed concrete
- Interior Finish: Red volcanic stone polishing, Paint finish on scratch cement
- Structural engineering: Eden Structure
- Mechanical engineering: SeongJi E&C
- Electrical engineering: I-eco ENG
- Construction: KR Construction
- Photography: Taxu Lee

Dosan Aloha
- Location: 654-8, Sinsa-dong, Gangnam-gu, Seoul, Republic of Korea
- Program: Commercial builiding
- Architects: Youngsoo Kim
- Site area: 273.7m²
- Building area: 161.0 m²
- Gross floor area: 730.48m²
- Building coverage ratio: 58.83%
- Floor area ratio: 199.30%
- Number of Levels: B1, 5F
- Structure: Reinforced concrete
- Exterior Finish: Sto Ventec R, Sto Signature, Stone
- Interior Finish: Fair-faced concrete
- Year completed: 2021
- Structural engineering: Eden Structure
- Mechanical engineering: Yirye E&C
- Electrical engineering: I-eco ENG
- Construction: Jium CM
- Photography: Joel Moritz

Brick Dunchon
- Location: 107-3, Dunchon-dong, Gangdong-gu, Seoul, Republic of Korea
- Program: House
- Architects: Youngsoo Kim
- Site area: 293m²
- Building area: 139.26 m²
- Gross floor area: 296.89m²
- Building coverage ratio: 47.53%
- Floor area ratio: 101.33%
- Number of Levels: 3F
- Structure: Reinforced concrete
- Exterior Finish: Brick
- Interior Finish: Gypsum Board W/ PAINT
- Structural engineering: Hyedam Structure
- Mechanical engineering: SeongJi E&C
- Electrical engineering: I-eco ENG
- Construction: Jium CM
- Photography: Joel Moritz

Lumen Yeonnam
- Location: 567-15, Yeonnam-dong, Mapo-gu, Seoul, Republic of Korea
- Program: Commercial building
- Architects: Youngsoo Kim
- Design team: Wonjoong Yang
- Site area: 194.4m²
- Building area: 116.25 m²
- Gross floor area: 468.64m²
- Building coverage ratio: 59.8 %
- Floor area ratio: 188.9%
- Number of Levels: B1, 4F
- Structure: Steel structure & Reinforced concrete
- Exterior Finish: Glass, Metal, Stucco
- Interior Finish: Stone Tile, Paint finish on scratch cement
- Structural engineering: Hyedam Structure
- Construction: Ziataek Construction
- Photography: Joel Moritz

Sara Hannam
- Location: 792-2, Hannam-dong, Yongsan-gu, Seoul, Republic of Korea
- Program: Commercial building
- Architects: Youngsoo Kim
- Site area: 123.4m²
- Building area: 73.10m²
- Gross floor area: 310.12m²
- Building coverage ratio: 59.23 %
- Floor area ratio: 198.36 %
- Number of Levels: B1, 5F
- Structure: Reinforced concrete
- Exterior Finish: Curtain-wall, Metal, Exposed concrete
- Interior Finish: Metal, Exposed concrete
- Year completed: 2021
- Structural engineering: Eden Structure
- Mechanical engineering: SeongJi E&C
- Electrical engineering: I-eco ENG
- Construction: Central Construction
- Photography: Joel Moritz, Jinbo Choi

SSamy Huam
- Location: 58-19, Huam-dong, Yongsan-gu, Seoul, Republic of Korea
- Program: House & Retail
- Architects: Youngsoo Kim
- Design team: Chaeyoon Song
- Site area: 111.4m²
- Building area: 63.8m²
- Gross floor area: 180.8m²
- Building coverage ratio: 57.27%
- Floor area ratio: 162.298%
- Number of Levels: 4F
- Structure: Reinforced concrete
- Exterior Finish: Brick
- Interior Finish: Gypsum Board W/ PAINT, Exposed Concrete
- Year completed: 2022
- Structural engineering: Hyedam Structure
- Mechanical engineering: SeongJi E&C
- Electrical engineering: I-eco ENG
- Construction: Sol-heim Construction
- Photography: Joel Moritz

Teyvat Tower
- Location: 325-25, Seogyo-dong, Mapo-gu, Seoul, Republic of Korea
- Program: Commercial building
- Architects: Youngsoo Kim
- Site area: 171.9m²
- Building area: 87.53m²
- Gross floor area: 487.92m²
- Building coverage ratio: 50.92%
- Floor area ratio: 228.64%
- Number of Levels: B1, 5F
- Structure: Reinforced concrete
- Exterior Finish: Stucco
- Interior Finish: Gypsum Board W/ PAINT, Metal
- Year completed : 2023
- Structural engineering: Hyedam Structure
- Construction: Jium CM
- Photography: Joel Moritz

리뷰 이성적 절제와 감성적 환희의 균형 박진호

건축디자인 행위는 기계적이라기보다는 본질적으로는 매우 인간적인 행위이다. 좋은 디자인은 정직하고 효과적인 방법으로 사용자와 소통하고 함께 호흡하게 한다. 얕은 지식이나 그럴듯한 속임수에 가까운 착취적 생산 전략으로서 만들어진 디자인이나 과대하게 치장된 디자인을 좋은 디자인이라 하기는 어렵다. 이런 디자인은 의사소통의 장애를 드러내는 주요한 요소이다. 과시욕, 지나치게 드러내고자 하는 욕망이 앞선 디자인은 결국 왜곡된 기능을 갖게 된다. 그렇지 않은 듯하지만, 많은 건축가가 자기 흥행 또는 넓은 의미의 건축업(business)이나 경제적 측면 등에 관심이 앞서 왜곡된 디자인을 양산하고 있음은 부정할 수 없다.

반면 김영수의 건축에는 과잉 디자인이 없다. 김영수는 으스대거나 값비싸게 치장한 듯한 허세를 담은, 자기 과시적이고 홍보를 위한 수단으로서 건축을 바라보지 않는다. 나르시시즘적 형태적 표현을 거부하는 대신 절제된 요소들을 질서 있게 구축하는 방식을 추구한다. 그는 과장 광고에 포장된 상품으로서가 아니라 완공된 건축물의 사용자를 중히 여기고 풍요롭게 해주는 건축을 추구한다. 건축물은 사용자가 원하는 방식대로 사용할 수 있어야 한다. 그는 건축가의 각본대로 사용하게끔 사용자를 유도하는 것이 아니라, 사용자 스스로 원하는 삶을 살 수 있도록 건축물이 일종의 배경 역할을 하는 것이라 인식하고 있다.

건축가는 디자인 행위를 통해 자신의 본성과 만나게 되고, 자신의 본성이 건축에 반영된다는 것을 안다. 김영수는 항상 평화롭고 조용하며 평정심을 잃지 않으면서 일관성 있게 자기 생각을 펼쳐간다. 차분하고 진지하며 냉정한 판단력과 자기 억제 능력이 강하긴 하지만 안정된 정서 속에서 자신의 방향성을 찾아간다는 점은 뚜렷하다. 과장되거나 지나치게 억지스러움은 찾아볼 수 없고 고요함의 무게감과 동시에 건축가로서 생명력을 보여주는 모습도 있다. 이러한 김영수의 품성은 건축 작업 과정이나 결과물에 그대로 반영된다. 요란하지 않고 군더더기 없이 깔끔하다. 멋져 보이려는 과장된 몸짓이나 억지스러움보다는 절제와 세련미가 돋보인다. 이는 그의 심리적 안정감과 더불어 내적 정신상태의 표출이다.

김영수의 작업 과정은 일상적 요소를 더욱 세심하게 변화시키면서 고된 작업에 창의성을 가미하고 상투적이며 관습적인 모습을 보다 참신한 존재로 바꾸는 것이라 할 수 있다. 단순한 작업 같지만 쉽지 않은 과정이다. 이는 김영수 내면의 굳건하거나 흔들리지 않는 방향성이나 신념의 결과다. 부차적으로 여기던 내용을 핵심 내용으로 치환하거나, 이러한 과정을 통해 건축의 인식에 심오한 변화나 진화를 가져오게 한다. 혁신은 확립된 질서의 재해석 과정을 통해 발전된다. 단순히 예외적 심미성을 탐구하거나 추구하는 것만으로 이루어 낼 수 없다.

Review
The Balance between Reasonable Moderation and Affective Delight

Jinho Park

Architectural design is essentially a very human, rather than mechanical, activity. A good design enables honest and effective ways of communing and keeping in tune with its users. It is hardly recognized as a design created with superficial knowledge or an exploitative production strategy tantamount to a specious trick, or as an excessively decorated design, which are indeed the main elements of revealing a kind of communication disorder. A show-off design, prioritizing an excessively exhibitionist desire, comes eventually to have distorted functions. Despite the specious looks, many architects are undeniably mass-producing distorted designs by prioritizing how to write their own success records or the interests in a broad sense of building business or economic aspects.

In contrast, Youngsoo Kim's architecture has no such excessive design. Kim does not regard architecture as a means of show-off promotion that is boastful or including pretensions like expensive decorations. He pursues an architecture that cherishes and enriches the lives of those supposed to live in completed buildings, and not as a commercial product packaged with exaggerated advertisements. Users should be able to use their buildings as they want. Kim does not lead the users to the architect's designed scenario, but recognizes his buildings as the backgrounds for them to live their own lives as they want.

The architect comes to meet his own nature through design activities, aware of how his nature is reflected in his architecture. Kim is always peaceful, quiet, losing no control of himself, and consistently unfolds his own thinking. Calm, serious, and cool-headed decision-making abilities strongly characterize his personality, but it is clear that he finds his own sense of direction in a stable sentiment. Any exaggerated or too far-fetched thinking cannot be found from his approaches, and even his vitality as an architect is shown along with the weight of quietness. This personality is reflected in his working processes or final products as it is, not boisterously, with no redundancy, neatly and efficiently. Rather than any exaggerated gesture or contrivance, moderation and refinement stand out. This is not only the mark of his mental stability, but also the articulation of his inner mentality.

Kim's working process can be said to change everyday elements more carefully, add creativity to laborious works, and transform stereotypical and conventional scenes into fresher ones. A simple work as it seems, it is not an easy process. This is the result of an unshaken directionality or belief in Kim's own inner mentality. What has been recognized secondary becomes reappraised by him as essential, or in this process, a profound change or evolution is brought on to the perception of architecture. Innovation develops through the process of reinterpreting the established order. It cannot take place only by exploring or pursuing an exceptional aestheticism.

자신의 공간구성 원리와 원칙의 바탕 위에 김영수의 건축 창자 과정은 공간을 찾아가고 발견해 가는 과정이다. 일상적 규칙들이 볼품없는 틀이나 획일적인 패러다임으로 김영수의 건축을 구속하고 있지도 않다. 그는 항상 공간을 먼저 생각한다. 평면으로 공간을 만들 수 없다. 공간을 먼저 상상하고 충분히 머릿속에서 다듬은 후, 평면을 만드는 것이다. 이러한 점에서 단면을 이용한 공간구성이라는 그의 디자인 사고방식은 아무리 강조해도 지나치지 않다. 도면을 그리는 행위도 공간을 발견하고 이해하며 드러내는 과정일 뿐 그리고 절차와 표현의 과정일 뿐, 도면 그 자체의 형식에 얽매이지 않는다. 이 과정에서 현재 진행되는 작업을 정확히 이해하고, 공간의 존재감을 상상하고 머릿속으로 구현하는 작업을 선행한다. 그리고나서 그 실행 방식을 찾아간다. 건축은 조각이나 드러나는 형태에 국한되지 않는다. 디자인 원리들은 눈에 보이지 않는 자신만의 도구로 활용될 뿐, 건축가 작업의 방점은 모두가 즐기는 감성적 공간이다. 공간은 차이를 만든다. 그러나 새로운 공간 사고는 저절로 나타나지 않는다. 다양한 시도를 통해 사고는 뿌리내릴 것이고, 그 깊이가 작품의 차이를 가름한다. 아무리 근사한 원리라도 그 자체로는 그다지 쓸모가 없다. 개념적 혁신과 사고의 변화가 필요하다. 김영수 건축의 미래 지향점이 있다면, 지속해서 공간을 자각하고 건축물 사용자의 삶을 풍요롭게 하고 영혼을 고양할 수 있을 만한 건축공간을 만들어 내야 한다는 것이다.

시대를 초월하여 사람들에게 감동을 주는 인류의 위대한 건축물에 숨어 있는 디자인 논리를 생각해 본다. 대성당 축조를 위해 중세 건축가들이 사용했던, 혹은 알베르티(Leon Battista Alberti)나 팔라디오(Andrea Palladio)와 같은 르네상스 건축가들이 가졌던 규율화된 원리와 이성적 논리에 기초하여 자기 창조의 논리로 발전시키려는 노력이 있었다. 근대에 와서는 구체적이거나 철저한 수학적 논리나 명확한 방법론은 아닐지라도 자신에게 맞는 방법론을 찾아간 루돌프 쉰들러(Rudolph M. Schindler)나 르 코르뷔지에(Le Corbusier)와 같은 건축가들이 있었다. 그들의 건축물은 무질서하게 형성된 공간이 아니라 내재한 원리를 통제함으로써 구축된 이성적 공간이면서 동시에 건축가의 상상이 결합된 결과물이다. 건축가의 지적 능력을 활용하여 건축물을 구축하는 과정에서 비례나 질서, 조화 등의 고전적 언어는 하나의 중요한 정신적 도구(mental tool)로 작용해 왔다.

건축적 질서의 이면에는 자신의 규칙과 방법론을 사용해서 디자인하고, 시행착오를 통해 경험을 축적하며 발전시켜 나가는 과정이 있다. 김영수는 고정관념이나 관습에 머무르지 않고 더 단순한 접근방식을 통해 규칙을 배우고 익히며 실험하면서 경험을 통해 자신만의 규칙을 만들어 나가고 있다. 루이스 칸 작업의 교훈에서처럼, 단순한 비율을 가진 사각형 같은 기본 형태에서 출발하여 대칭, 분절, 반복, 리듬 등의 고전적 기법을 통해 질서를 구현한다. 원리와 원칙은 단순하나 그러한 원리와 원칙을 이용하는 사람의 능력은 다양하다. 이성적 논리로 시작하되 다양한 아이디어와 개념을 적용하면서 디자인은 점점 더 세밀해진다. 시행착오를 겪으면서 첫 단계의 추측보다 점점 더 나은 답을 얻으면서 나아간다. 인간의 오랜 지성이 진화하는 모습이다. 단순한 것에서 출발하여

Based on his own spatial organization principles, Kim's architectural creation process is to go in quest of and discover spaces. Everyday rules are not constraining his architecture with any ugly frame or uniform paradigm. He always considers space first of all. Floor plans are not the trigger of space. One should first imagine a space, elaborate it in your head, and articulate it into floor plans. In this respect, his design thinking with spatial composition in section cannot be overemphasized at all. The activity of drawing is just a process of discovering, understanding, and revealing space. It is just a procedure and process of expression, which is not constrained by the form of drawing itself. In this process, one understands the on-going work correctly, imagines the presence of space, and prefigures the work of embodying it in one's head. Then, one goes in quest of how to implement it. Architecture is not limited to sculpture or revealed forms. Design principles are just utilized as one's own invisible tools, while the architect's work focuses on an emotional space where everyone can enjoy. Space creates differences, but a new spatial thinking does not occur automatically. Thinking will be rooted through various attempts, and its depth determines how the work differs. No matter how awesome, the principle itself is not quite useful. We need conceptual innovation and the change of thinking. If Kim's architecture should have a future destination, it would be to keep conscious of space and create an architectural space that enriches the users' lives and enhances their souls.

I think about the design logic hidden behind the humankind's great buildings that impress people in a timeless way. There were such efforts as to develop one's own creative principles based on the disciplined principles and rational logics used by medieval architects for the construction of cathedrals or by such Renaissance architects as Leon Battista Alberti or Andrea Palladio. In the modern age were such architects as Rudolph M. Schindler and Le Corbusier that went in quest of methodologies suitable for them although they were not specific or thorough mathematical logics or clear methodologies. Their buildings were not disorderly created spaces, but the products of combining the architect's imaginations with the rational spaces constructed by controlling their immanent principles. In the process of constructing buildings with the architect's intellectual abilities, the classical language such as proportion, order, and harmony has served as an important mental tool.

Behind the architectural order is a process in which one designs with one's own rules and methodologies, building and developing experiences through trials and errors. Kim is creating his own rules through experiences by learning, mastering, and experimenting such rules in a simpler approach, rather than staying at stereotypes or conventions. Like the lessons from Louis Khan's works, he starts from such basic shapes as simply-proportioned rectangles and embodies the order through such classical techniques as symmetry, articulation, and rhythm. Simple as they are, these principles are used by the one who has diverse abilities. Beginning with a rational logic, he applies various ideas and concepts so as to elaborate the design. Experiencing trials and errors, he proceeds with tentative answers better than the first speculation. It is in the same manner that the human intellect has evolved for a long time, through

적응하고 배우는 능력 말이다. 배움은 실험을 통해 세상과 교감하면서 얻게 된다. 합리성조차도 어찌 보면 그에게는 경험의 과정이다. 다양한 지식을 습득하고, 축적된 지식과 자료를 토대로 실무에 응용하고, 시행착오를 통해 배우는 것이다. 실패와 피드백을 받아들이며 지속적인 학습을 통해 자기 능력을 향상하면서, 그러면서도 타협을 통해 세상과 소통하고 공유하는 언어를 만들어 나가야 한다. 건축에 완벽한 모델은 없다. 자기 주도적 건축 모델을 찾아가는 과정은 젊은 건축가에게 어찌 보면 숙명이다.

김영수에게 이러한 접근방식을 이끈 건 교육의 힘이다. 대학원 교육과정에서 조영이론에 대한 탐구나 개인 연구를 통해 과거 위대한 건축의 복잡성에 내재한 원리를 이해하고 자신의 건축에 적용해 보려는 일련의 노력이 중요한 밑거름이 되었으리라 판단한다. 규칙을 통한 집합적인 조직과 그 변화의 법칙을 이해하는 것은 하나의 도전이고 평생의 과제일지 모르지만, 좋은 건축가로 성장할 수 있는 토양이 될 것임에는 틀림없다. 건축은 기본적 원리를 주무르고 변형시키는 건축가의 능력과 솜씨에 따라 다양한 결과를 얻어내는 창의적 작업이다. 논리만 따르는 비인간적인 작업이 아니라 빈 땅에 새로운 형상을 부여하여 미적 감동을 자아내는 과정이다. 이러한 과정으로 만들어진 작품은 최종적이라기보다 미래에 더 발전할 수 있는 사고 진화 과정의 일부분이라 할 수 있다.

동시대를 사는 사람들과 어울리면서 그들과 함께 건축하는 패턴이 있다. 이 과정은 수월하지만, 평범한 디자인만을 양산해내기 쉽다. 보통 개인보다는 집단적 패턴으로 이 과정에서 하나의 흐름을 형성하게 되는데 김영수는 이러한 흐름을 따르지 않는다. 김영수의 건축은 자신만의 색깔이 있다. 자신의 철학과 방법에 근거하여 디자인을 만들어나가면서도 특정 스타일을 추구하거나 다양한 요소들을 빌려 뒤섞는 광적인 방식을 추구하지 않는다. 포스트 모더니스트만큼이나 무작위적이거나 광적인 결합을 자제하고 꽤 단순한 원칙을 따른다. 하지만 상당히 놀라운 결과를 만들어낸다. 다이아몬드의 광택이 아름다운 이유는 그 구성 원자가 빛나기 때문이 아니라 원자들의 조직이 특별해서 더 빛나고 아름답다. 누구나 같은 건축 요소, 유사한 언어를 사용하지만 중요한 것은 그 요소가 아니라 그것을 구성하는 질서이다. 물론 사소한 조합이라는 우연한 결과를 만들 수도 있지만, 이러한 질서에 대한 고민이 근본적이고 반복되어 디자인에 적용된다면, 더욱더 김영수만의 독특하고 깊이 있는 건축을 만들 수 있을 것이라 확신한다.

이러한 김영수의 건축적 내면은 작품 속에서도 디자인 언어로 뚜렷이 나타나고 있다. 공간 형태의 구축에 있어 절제된 언어의 조합, 프로젝트마다 다양한 조건에서의 질서 구축, 건축물의 완성도를 높이기 위한 디테일에 대한 고민, 재료의 물성과 그 활용 가능성 등을 다양하게 모색함으로써 자신의 건축적 방향성을 찾아가고 있다.

김영수의 건축은 간단한 규칙, 리듬, 그리고 기하학적 법칙을 적용함으로써 질서의 기초를 구성한다. 주택 나지요네의 구성법을 보면 프랭크 로이드 라이트(Frank Lloyd Wright)의 주택 배열 원칙 중 주개념인 바람개비(pin-wheel) 형태의 기하학적 평면 구성을 떠오르게 한다. 또한 루돌프 쉰들러의 포페노(Popenoe) 주택 배열에서나 기타 디자인에서도 주요한 방법론으로 사용되는 방식들이 있다. 동서남북으로 뻗은 각 방에 특

one's abilities to adapt oneself to and learn something by starting from simple things. Learning occurs by communicating with the world through experiments. Even rationality is a process of experience for him: acquiring various kinds of knowledge, applying them to practice based on the accumulated knowledge and data, and learning through trials and errors. He should accept failures and feedbacks, improving his own abilities through ceaseless studies, while creating a language that can be shared to communicate with the world through compromises. There is no perfect model in architecture. In a way, the process of seeking for a self-directed architectural model is the destiny for a young architect.

What has driven Kim's approach in this way is the power of education. I believe a series of efforts he made, to understand the principles inherent in the complexity of great historical architecture by studying building theories in graduate school or personally, would have been an important foundation. Understanding the rule-based organization of collective entities and the principles of their transformations might be a challenging or even lifelong task, it must surely be the ground to grow as a good architect. Architecture is a creative operation with various outcomes entailed by the architect's abilities and skills that deal with and transform its basic principles. It is not a non-human operation following just logics, but a process of giving a new figure to the empty site so as to move our minds aesthetically. The works created in this way can be seen as part of a thought evolution process which will be able to develop in future rather than be complete now.

There is a pattern of contemporaneity in which peer architects are mingling and working together. This process is easy, but likely to produce only ordinary designs. It usually occurs in a collective, rather than individual, pattern that forms a kind of school. Kim does not follow this, however. His architecture has its own color. His designs are created on the ground of his philosophy and methodology, but not pursue a specific style or a crazy combination of mixing borrowed elements diversely. Abstaining from random or crazy postmodernist combinations, he follows quite simple principles, still creating considerably wonderful results. It is not that the individual shiny atoms constituting a diamond make its luster beautiful, but that their organization makes it shinier and beautiful. Although everyone uses the same architectural elements and similar language, what is important is not such elements but the order of their composition. Accidental results could be made by trivial combinations, but I believe that if some ruminations upon this order are fundamentally conducted and repetitively applied to designs, there can be occur Kim's own architecture that is all the more unique and profound.

This inner mentality of Kim's architecture is also clearly found in his works as a design language, finding his own architectural direction by combining controlled "paroles" in constructing spatial forms, building an order in various conditions by projects, considering details to improve the quality of buildings, and exploring materiality and its possibilities for many different uses.

Kim's architecture organizes the foundation of order by applying simple rules, rhythms, and geometrical principles. Looking how the house Nasillonner

별한 조망(view), 나아가 시선의 방향성을 뚜렷이 부여하고 있다. 또 중정 형식의 마당을 제공함으로써 제주도라는 지역 기후적 특성에 최적화된 평면과 특별한 내외공간의 소통을 시도하고 있다. 이러한 접근 방식은 매스의 분절과 함께 각각의 공간에서 특별한 외부 조망과 자연광을 즐길 수 있도록 계획된 브리크둔촌에도 그대로 적용된다. 브리크둔촌은 층별로 각 방과 외부 테라스로 연결된 공간에 조망 요소를 담고 있으며, 주 공간 사이사이의 복도에도 다양한 외부 조망을 제공하고 있다.

특히 도시 내 상업적 건물에서의 입체와 리듬감은 많은 무질서한 교란과 간섭에도 김영수의 잠재적인 능력을 충분히 드러낼 수 있을 정도로 잘 정리·정돈되어 있다. 상업가로에서 일조선 규정의 결과라고 치부하기에는 매스 형상의 다룸이 치밀하고 세부 디테일의 기지(wit)를 볼 수 있다. 사라한남의 경우 작은 근린생활시설의 특성상 많은 디자인 요소를 사용할 수 없는 상황에서도 일상의 작은 뒷골목에 어떤 특별함을 부여하고 있다. 수직적 리듬과 규칙이 만드는 상업가로의 모습은 주변 경관 변화의 씨앗이 되고 있다. 특히 도산 알로하의 경우 입면 디자인을 자세히 보면, 단순하지만 세밀한 디테일의 변화를 통해 비례감이나 리듬감을 규정하고 있다. 이와 달리 연남동 대수선 건물(루민연남)에서는 층마다 서로 다른 길이의 다소 트렌디(trendy)한 철과 유리를 수평 요소로 사용하여 거리의 리듬감을 배가하고 있다.

많은 건축가가 자기 작품의 디테일을 세밀하게 계획하고 해결해 보려는 시간이 짧아지고 있다. 자신의 시간 중 90% 이상을 디테일 해결에 사용한다는 프랭크 게리(Frank O. Gehry)의 말처럼 디테일의 차이가 건축물의 품격을 결정한다고 생각한다. 건축 작업의 초기부터 이러한 디테일에 관한 깊이 있는 관심은 좋은 시작이라 볼 수 있다.

'젊은' 건축가로서 건축 재료의 물성, 기능적 특성뿐 아니라 건축공간에 미치는 질감(texture) 등을 고민하면서 재료별 사용 방식이나 각 재료의 특성에 맞는 건축디자인 조합을 위한 합리적 방안을 실험하는 것은 당연한 실무적 배움 과정이다. <수리움>에서의 붉은 송이석 노출콘크리트, <나지요네>에서는 회색 노출콘크리트, <도산 알로하>와 <사라한남>에서는 금속 외피, <브리크둔촌>의 벽돌, 그리고 <가온누리>에서의 목재 외장재 등 일련의 프로젝트에서 같은 재료를 사용하기보다는 프로젝트의 성격과 목적에 맞게 다양한 재료를 시도하려는 김영수의 고민을 엿볼 수 있다. 프로젝트마다 다른 재료를 사용해 보는 것은 단발성 실험이 아니라 지속적이고 상호 관련된 재료를 찾아가는 과정이다. 건축디자인은 우연의 일시적 결과물이 아니라 배움의 결정체이고 맹목적인 양식적 편견이나 추종에서 탈피하여 주도적인 힘을 만들어 가는 과정이다. 여기에 다양한 재료, 시공 방법, 그리고 구조 등에 관한 기술적 지식과 현장경험을 갖춘다면 자신의 건축을 구현하는 데 배가된 능력을 발휘할 것이다.

자신의 체화된 지식과 경험을 몇몇 건축 작품에 충분히 담기에는 지금까지 걸어온 길이 아직은 짧다. 작품들에 자신만의 색깔을 담아내기에도 그 규모가 너무 작아서 한계가 있다. 연배 있는 건축가들에 비해 자신의 건축 철학과 상상력을 펼쳐볼 기회도 많이 없었을 것이다. 그러나 재능도 있고 실력도 있는 건축가가 부단히 노력하면서 창의적인

was organized, I am reminded of Frank Lloyd Wright's geometrical floor planning in a pin-wheel shape, one of the main concepts of his house layout principles. Also, Nasillonner reveals several aspects used as the main methods in Rudolph Schindler's layout of Popenoe cabin or other designs. Each of the rooms stretching in the four cardinal directions is provided with a special view and even the directionality of sight clearly. Besides, the courtyard is provided to attempt a special communication between the inside and outside with the floor planning optimized to the local climate of the Jeju island. This approach also applies to Brick Dunchon, which was designed to make it possible to enjoy daylight and a special view to the outside in each space along with articulated massing. Brick Dunchon provides scenic elements by floors at every connection between each room and the outdoor terrace, and also various views to the outside at the corridors between spaces.

Particularly, its volume and rhythm seen amongst the other urban commercial buildings are so well-organized as to reveal Kim's potential abilities enough despite many disorderly disturbances and interferences. The massing is so elaborately formed with the wit of details that it could hardly be dismissed as a result of setback regulations from the commercial street. As for Sara Hannam, it provides some uniqueness for the everyday small back street even without the chance of using many design elements due to the characteristics of a small neighborhood living facility. The commercial street scene created by the vertical rhythm and rules is serving as the seed of changing the surrounding scenes. For Aloha Dosan, particularly, its facade design shows how the sense of proportion or rhythm is prescribed through the change of simple but minute details. Unlike this project, the remodeled Lumen Yeonnam shows how the street's rhythmical sense is multiplied by the horizontal and somewhat trendy use of steel and glass with different lengths by floors.

Many architects are increasingly experiencing the lack of time spent on working out the elaborate design of their working details. As Frank O. Gehry said he spent more than 90% of his time on working out details, I think the difference of details determines the quality of buildings. This profound interest in details from the early phase of an architectural project can be seen as a good beginning.

It is a matter of course for a "young" architect's practical studies to mull over material characteristics such as materiality itself, its functional characteristics, and its texture perceived within architectural space, experimenting with reasonable measures to combine architectural designs fit for the peculiar treatments and characteristics of each material. Kim's ruminations to attempt various materials fit for the character and purpose of each project rather than same materials can be glimpsed through a series of projects such as: *Surium* with red volcanic-stone-colored exposed concrete; *Nasillonner* with grey exposed concrete; *Aloha Dosan* and *Sarah Hannam* with metal envelopes; *Brick Dunchon* with bricks; and *Gaon Ruri* with wooden cladding. Using various materials for different projects is not a one-off experiment but a continuous searching process to discover interrelated materials. An architectural design is not a temporary outcome of chance, but a crystallization of studies and a process

모험을 주저하지 않는다면 기회는 오기 마련이다. 아직 젊지만 이 정도 작품활동을 하는 것을 보면 근미래 전개될 김영수의 작업에서 더 큰 감동과 가치를 주는 작품이 나올 것이라 예견하게 된다. 나아가 창작의 열정만으로 이루는 건축이 아니라 건축의 사회적, 역사적, 예술적 가치의 영역까지 담을 수 있는 의미 있는 활동을 이어갔으면 한다. 그 과정에 내재한 자신만의 접근 방식이나 디자인 원리가 반복된 건축은 이후 김영수만의 특화된 건축세계를 드러내는 힘이 될 것이다.

박진호
박진호는 미국 하와이주립대학교 건축대학에서 부교수(종신교수)를 역임했다. 현재는 인하대학교 건축학과 교수로 재직 중이다. 전공 관심 분야는 건축디자인 방법론, 이론 및 역사, 건축계획 및 설계, 디지털 디자인 및 로보틱스 제작, 디자인 컴퓨테이션 등이다. 『Nexus Network Journal』(Birkhauser, Springer Nature) (A&HCI)의 편집위원을 역임했고, 현재는 『Open House International』(SSCI and A&HCI) 편집위원을 거쳐 객원 편집자(Contributing Editor)로 활동하고 있다. 저서로는 『Architectural and Urban Subsymmetries』(Birkhauser, Springer Nature, 2022), 『Graft in Architecture: Recreating Spaces』(Mulgrave: Images Publishing, 2013), 『Designing the Ecocity-in-the-Sky』(Mulgrave: Images Publishing, 2014) 등이 있다.

of creating an initiating power escaping from any blind assumption or pursuit for an existing style. If equipped here with the technical knowledge and field experiences on various materials, construction methods, and structures, the architect would be able to display a doubled capability to materialize his own architecture.

The way he has walked through to date is yet to be long enough to reflect his embodied knowledge and experiences in some architectural works. Perhaps he would have had not so many chances to unfold his architectural philosophy and imaginations compared to more seasoned architects. Be that as it may, there comes a certain chance for a talented and capable architect who ceaselessly endeavors to do creative adventures without hesitation. As Kim is still young but already showing this degree of competent works, I look forward to his near-future works providing more impressions and values. Furthermore, I hope he will not only depend on creative passions for architecture but also continue meaningful activities that can reflect its social, historical, and artistic values. Such an architecture as to reflect his own approach or design principles repetitively will reveal his own specialized architectural world thereafter.

Jinho Park
Jinho Park served as an associate (tenured) professor at the University of Hawaii School of Architecture, and now is serving as a professor at the Inha University Department of Architecture. His major areas of concern include architectural design methodology, theory and history, architectural planning and design, digital design and robotic fabrication, and design computation. He worked as an editor for *Nexus Network Journal* (Birkhauser, Springer Nature) (A&HCI) and *Open House International* (SSCI and A&HCI), and now is active as a contributing editor for the latter journal. His authored books include *Architectural and Urban Subsymmetries* (Birkhauser, Springer Nature, 2022), *Graft in Architecture: Recreating Spaces* (Mulgrave: Images Publishing, 2013), and *Designing the Ecocity-in-the-Sky* (Mulgrave: Images Publishing, 2014).

AGIT STUDIO
아지트 스튜디오

Jamin Seo
서자민

서자민은 2017년 아지트스튜디오 건축사사무소를 설립하고 대표 건축가로 작업을 진행하고 있다. 연세대학교 건축학과와 UPENN 건축대학원에서 학위를 취득하고 삼우종합건축사사무소 및 원오원 아키텍스에서 실무를 쌓았다. 현재 고려대학교 건축학과의 설계 스튜디오를 맡고 있다. 아지트 스튜디오의 건축은 좋은 질문을 만드는 것과 상황에 대한 고유한 해석에서 출발한다. 온기를 머금지만 차가움과 낯섦을 유지하는 건축이다. 이를 위한 건축적 집요함을 만드는 곳이 아지트이다. 건축가 허근일이 객원 파트너로 작업과 토론에 참여하고 있다.

Jamin Seo founded AGIT STUDIO in 2017 and has been working as a principal architect. She received her degrees from the Department of Architecture of Yonsei University and the University of Pennsylvania School of Design, and built her career with Samoo Architects and One O One Architects. She is currently in charge of a design studio of the Department of Architecture at Korea University. AGIT STUDIO's architecture starts with developing good questions and producing a unique interpretation of a given situation. It is about architecture that retains warmth while maintaining a sense of coldness and unfamiliarity. AGIT is the place for nurturing architectural tenacity to achieve this goal. Architect Guenil Huh participates in design work and discussion as a visiting partner.

Attitude of AGIT STUDIO

태도; Attitude

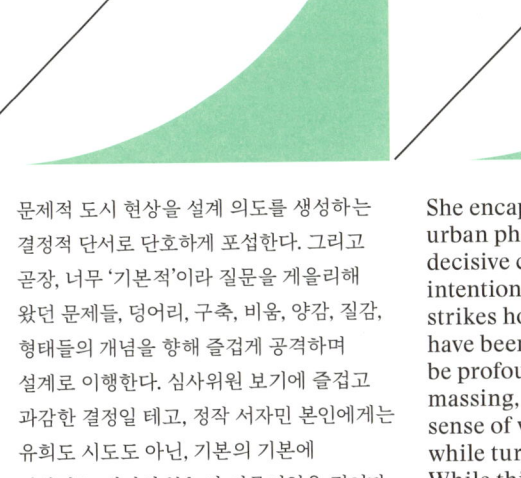

문제적 도시 현상을 설계 의도를 생성하는 결정적 단서로 단호하게 포섭한다. 그리고 곧장, 너무 '기본적'이라 질문을 게을리해 왔던 문제들, 덩어리, 구축, 비움, 양감, 질감, 형태들의 개념을 향해 즐겁게 공격하며 설계로 이행한다. 심사위원 보기에 즐겁고 과감한 결정일 테고, 정작 서자민 본인에게는 유희도 시도도 아닌, 기본의 기본에 천착하는 내면의 본능적 씨름이었을 것이다. 자신의 질문에 대한 명료한 입장을 제시하는 작업 과정은 젊은 건축가의 고유한 내러티브를 넘어 기성 건축계의 어쩌면 빈궁한 담론에 반한 원초적 물음을 던진다.

— 심사평 중

She encapsulates a problematic urban phenomenon sternly with a decisive clue that creates a design intention. Straight away, she enjoyably strikes home of the issues that have been too "basic" concepts to be profoundly questioned, such as massing, construction, emptying, sense of volume, texture, and forms, while turning them into design. While this was probably seen by the jury as a joyful and bold decision, it would have been neither a play nor an attempt for Seo herself but just her instinctive inner struggle delving into the very basics. The process of presenting a clear position on her own questions goes beyond the young architect's unique narrative, posing fundamental questions that challenge the impoverished discourse of the established architectural community.

— From Jury's comment

텍토닉', 확장	206
내러티브	194
못생김	184
고집	174
덩어리	162
태도	160

Attitude	*161*
Mass(Deongeori)	*163*
Persistence	*175*
Ugliness	*185*
Narrative	*195*
Tectonic, Expansion	*207*

태도

'태도'라는 말은 어떤 상황에 대한 마음가짐, 모양새, 입장 등이 될 수 있겠다. 조금의 여유치를 가진 이 단어에 기대어 말하고 싶은 것은, 복잡하고 긴 과정 위에 있는 '건축행위'에서 건축가에게 시작점이나 기준이 될 수 있는 어떤 지점에 대한 이야기일 것이다. 우리의 생각과 마음가짐, 곧 '태도'라 할 수 있는 몇 가지 이야기로 아지트 스튜디오 프로젝트들을 엮어보기로 한다.

의문을 품거나 호기심을 가지고 관찰하는 행위는 일상의 장소에서, 공간에서, 익숙한 도시에서 그리고 새로운 여행지에서도 이어진다. 성장 과정, 교육 과정, 실무 과정 등 각각의 과정 자체와 그 사이를 채워온 다양한 경험은 모두 의미 있는 내외부적 자극이었다. 이러한 경험에서 나온 관심과 생각은 프로젝트를 만났을 때 고유한 질문을 만든다. 이 질문을 바탕으로 '무엇을 하겠다'라는 의도는 건축작업의 중심이 된다.

언제나 완성되지 않고 진행 중인 이 상태를 무 자르듯 정리하기는 어렵다. 그러나 태도는 우리가 만든 건축물들을 느슨하게 연결 짓는 매듭 혹은 기저 어딘가에 있을 원동력을 짐작하게 할 것 같다. 나는 영화를 볼 때, 때때로 감독의 필모그래피나 평소 관심사들을 찾아보기도 한다. 각 작품의 의미와 완성 너머에 변화하며 이어지는 창작자의 생각을 읽는 것은 영화를 보는 데 재미와 도움을 주기 때문이다. 각 시기에서 우리의 생각, 경험, 열망은 태도로서 곧 건축물이 되었고, 태도는 지금도 변화하며 나아가고 있는 중이다.

Attitude

The word "attitude" implies a frame of mind, appearance, or stance toward a certain situation. What I want to say, relying on this word, is a story about a certain point that can be a starting point or standard for an architect in the "act of architecture" that is a complex and lengthy process. Compiling Agit Studio's projects, several stories will be told here about our thoughts and frame of mind, that is to say, "attitude."

The act of asking questions or observing with curiosity continues in everyday places and spaces, familiar cities, and even new travel destinations. Each process of growth, education, and professional work itself as well as the various life experiences in between them are all meaningful internal and external stimulations. The interests and thoughts developed from these experiences generate unique questions when faced with a project. Based on these questions, the intent of "what to do" becomes the center of architectural work.

This state, always unfinished and in progress, is not cut and dried. However, the attitude allows us to guess the knot that loosely connect the buildings we make or the driving force that lies somewhere underneath. When I watch a movie, I sometimes look up the director's filmography or personal interests. This is because reading how the creator's thoughts have changed beyond the meaning and completion of each work is interesting and helpful in enjoying the movie itself. Our thoughts, experiences, and aspirations in each period were our attitudes, which soon became buildings. These attitudes are still changing and in progress.

덩어리

지하철은 산티아고 도심을 지나 산 호아킨(San Joaquin)역에 다다르고 있었다. 빨간 비니를 덮어쓰고 이어폰을 낀 남자가 기댄 창문 밖으로 갑작스레 덩어리가 드러났다. 놀랐던 것은 예상보다 크게 주변을 압도하고 있는 규모였다. 그 덩어리는 고가 위를 달리는 지하철에서부터 인식되는 도시적 스케일이었다는 것을 근처에 다다라서야 알게 되었다. 알레한드로 아라베나(Alejandro Aravena)가 설계한 칠레의 카톨리카 대학 <UC 이노베이션 센터>를 찾아가는 길이었다. 잔뜩 기대를 안고 도착한 곳에서 나는 한참 동안 거의 뛰다시피 다니며 건물을 구경했던 것 같다. 페루에서 시작한 거친 대륙 여행길이 국경을 지나 칠레에 진입했고, 그 국토의 길이만큼이나 위도를 따라 띄엄띄엄 늘어선 도시들을 차례로 건너는 와중이었다.

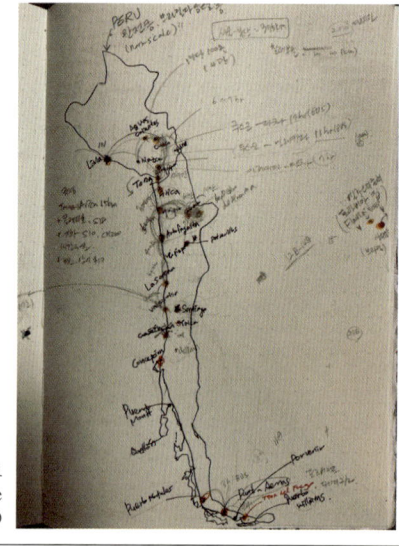

남미기행 루트
Travel route
ⓒAGIT STUDIO

2017년 여름, 터질 듯 빡빡하게 이어지는 하루하루의 연속 위에서 거짓말처럼, 그리고 마침내 찾아 떠난 곳은 남미였다. 당시 나는 정식으로 아지트 스튜디오를 만들어 독립하기로 마음먹고 퇴사한 직후였고, 파트너인 허근일은 김태수 펠로우십(2016)[1]을 수상하고 한 해가 지나도록 휴가를 미룬 탓에 그 혜택의 기회가 사라질 시점이었다. 남미여행은 그렇게 조금은 다른 명분과 조건을 가진 두 건축가가 떠난 건축기행이었다. 당시 우리 각자에게는 질문과 갈증이 쌓여있었던 것 같다. 수상자의 의견에 앞서 나에 의해 정해진 여행지에 불평을 듣기는 했지만, 남미를 가야하는 데 이견은 없었다. 나는 안데스산맥을 경계로 태평양과 닿은 페루와 칠레의 최남단 끝까지 약 5천km를 종단하는 루트를 제안했고 허근일은 그것을 육로로만 이동해보자고 덧붙였다. 뭐라도 덜어내야 하지 않냐는 입장과 도무지 덜어낼 것이 없다는 입장이 투덕거렸지만 남미의 '토레스 델 파이네'는 당신도 꼭 가보고 싶은 곳이라고 하셨던 김태수 선생님의 메일 덕에

---

[1] 김태수 해외건축여행 장학제:
T. S. Kim Architectural Fellowship Foundation은 1992년부터 매년 한 명의 젊은 건축가를 선정해 세계 건축여행을 지원하고 있다.

Mass(Deongeori)

The subway train had passed through downtown Santiago and was approaching San Joaquin Station. A man wearing earphones and a red beanie is leaning against a window. Outside that window, a mass suddenly appeared. At first, I was surprised at how it clearly overwhelmed its surroundings on a massive scale, which was not what I had expected. It was only when I got closer that I realized that the mass was of an urban scale recognized from a subway running on an overpass. I was on my way to *UC Innovation Center* designed by Alejandro Aravena at the Pontifical Catholic University of Chile. I had high expectations for the building, and when I arrived there, I spent a significant amount of time looking around, almost running here and there. Our trip across South America started in Peru. The journey was tough, and we crossed the border and entered Chile. We passed by city after city scattered across different latitudes of the elongated country.

In the summer of 2017, on the verge of burnout from spending each day intensely, I packed up my things and headed to South America. It was right after I left my former employer to work independently, having made up my mind to open Agit Studio. My business partner, Guenil Huh, who won the T. S. Kim Architectural Fellowship(2016),[1] had postponed going on an architectural journey for nearly a year and was about to lose his eligibility. The trip to South America was a journey of two architects traveling for different reasons and under different conditions. Looking back, I think each of us had questions and thirst piled up inside. The fellowship recipient complained that I had chosen the travel destinations without asking his opinion, but there was no disagreement that we should go to South America. I proposed a journey spanning approximately 5,000 km along the Andes Mountains bordering the Pacific Ocean, from Peru to the southernmost tip of Chile. Guenil Huh proposed that we travel the route only by land. He

---

1   T. S. Kim Overseas Architectural Travel Scholarship: Since 1992, the T. S. Kim Architectural Fellowship Foundation has selected one young architect every year to support his international architectural travel.

욕심냈던 모든 루트를 지킬 수 있었다.

우리가 대학에서 교육을 받고, 이후 실무에서도 학습하게 되는 '참조가 될 만한 건축'은 유럽과 미국, 그리고 일본 등에 위치해 있는 경우가 대부분이다. 건축이라는 것이 자본과 깊이 연관되어 있고, 잘 갖추어진 사회 시스템과 기술이 바탕이 되어야 정교함과 높은 완성도를 실현시킬 수 있기 때문이기도 하다. 하지만 자본과 예리한 기술력이 부족한 나라에서도 좋은 건축 작업은 일어나고 있다는 사실은 나에게 흥미롭고, 그 과정과 동력에 대한 상당한 호기심을 자아내는 일이었다. 더욱이 중남미의 몇몇 국가와 지역은 현대건축의 영역에서 이미 주요 관심의 자리를 차지하고 있지 않나. 그것을 직접 보고 싶었다.

특히 목적지로 안데스 서쪽을 택한 건 근대건축의 영향권을 다소 비켜 간 토양과 그 지역의 건축이 궁금했기 때문이다. 남아메리카 땅 특유의 러프함, 그리고 거기에서 오는 아우라가 건축에서 느껴질 수 있기를 기대했다. <UC 이노베이션 센터>에서 뛸 듯 좋아했던 건 보고 싶었던 단서를 찾았기 때문이었다. 매스(덩어리)만으로 존재감을 충분히 증명하는 것, 어떤 디테일의 놀라움이 발견되는 것이 아니라 그저 건물로서 좀 더 본질적인 힘을 발하는 것. 거친 콘크리트 표면에 못구멍이 드러나 있는 곳도 많았고, 창과 금속들의 접합부 등 '섬세'란 말을 꺼내지도 못할 부분들도 보였지만, 그것은 우리가 즐거움을 누리는 데 아무런 방해가 되지 않았다. 중요한 건 현대건축에서도 분명 그것을 가능하게 하는 지점이 있을 것이라는 질문에 대한 확인이었다.

UC 이노베이션 센터
Innovation Center UC
©AGIT STUDIO

argued that we should filter out some destinations while I counter-argued that nothing should be taken out. I was able to keep all destinations in the travel route thanks to an email from Taesoo Kim, who said that Torres del Paine in South America is in his bucket list of places to visit.

Most of architecture that we study in school and "worthy of referencing" in doing our work is located in Europe, the United States, and Japan. This is because architecture is deeply related to capital, and its sophistication and high degree of perfection can only be achieved when supported by well-equipped social systems and technologies. However, the fact that good architecture was produced even in countries lacking capital and refined technique was interesting and I was extremely intrigued by the process and dynamics. Moreover, several countries and regions in Latin America were already attracting attention with their modern architecture that reflects their respective environments. I wanted to see it for myself.

In particular, the reason I chose the west side of Andes as our destination was because I was curious about the soil and architecture of that area, which was somewhat immune to the influence of modern architecture. I hoped that the unique roughness of South American soil and the aura that comes from it could be felt in the architecture. The reason I was excited at *UC Innovation Center* was because I found the clues I was looking for. In other words, the mass alone was sufficient proof of its presence. It was not about discovering any surprises in the details, but simply about emanating the essential power of a building itself. There were many nail holes exposed on the rough concrete surface, and the joints between the window and metal looked sloppy or not delicate enough, but those things didn't whatsoever disturb us from enjoying the building. What is important is the question of the point of modern architecture that allows that experience.

Agit Studio

아지트 스튜디오

<프로젝트 재해석>은 붉은 덩어리를 가지고 있다. 40년 연식의 두 상가 건축물 사이를 이어주는 신설 코어이자 기존 건물 위에 올라간 증축 부분이다. 클라이언트의 요청 사항과 제시된 프로젝트의 목적은 명료했다. 노후화된 기존 건물에 엘리베이터 설치하고 두 건물에 각각 있던 계단실을 한 곳으로 합쳐 추가 면적을 확보할 것, 외장재 교체 등 상가건물의 임대수익 향상과 유지관리 편의를 위한 리모델링을 하는 것. 분명하고 구체적인 요구사항이었던만큼 다른 것에서 차별적인 접근을 하는 것은 여러모로 합리적이지 않았다. 그러나 우리는 이 건물이 단지 보수를 마친 증축 건물이 되기보다 노후화된 상가건물을 리모델링하는 방식에 대한 의도와 메시지를 드러내고 싶었다. 그렇게 하기 위해서는 '당연한' 요구 조건에 건축가로서의 재해석이 필요하다고 여겼다. 두 건물이 놓인 대상지는 한때 부산의 잘 나가던 중심 상권에 위치하고 도심 상권의 보편적인 모습을 가지고 있다. 지금은 다소 잠잠해진 구도심 대학가이지만 여전히 중심가로이고, 앞으로는 이 두 건물이 하나로 합쳐져 두 배의 규모가 될 것이다. 이런 존재감을 건물의 수리 보수만으로 끝낼 수는 없다고 생각했다.

프로젝트 재해석
Project Re-interpret
ⓒKyungsub Shin

Attitude of AGIT STUDIO

Project Re-interpret has a red mass. It is a new core that connects two 40-year-old commercial buildings as well as an expansion built on top of the existing building. The client's requirement was clear and specific, which is to install an elevator, merge staircases for additional space, and remodel the commercial building for higher rental income and convenient maintenance. As the requirement was clear and specific, it was not reasonable in many ways to approach the project differently. However, we wanted to showcase the intent and message of how an aging commercial building was remodeled rather than showing the results of a building repair followed by expansion. To do so, as architects, we believed that it was necessary to reinterpret the "straightforward requirements" of the client. The site that the two buildings occupy is located in the center of a commercial area which was once a symbol of the hustle and bustle of Busan. The area possesses the typical appearance of a downtown commercial district. Although it is now a somewhat quiet university district in the old city center, it is still a central area. When the two buildings are combined into one, it will double in scale.

우리는 디자인 의도나 개념을 다른 데서 끌어오기보다 요구사항에 따라 새롭게 변하는 건물의 변화를 드러내는 데에 집중했다. 그리고 그 행위 자체를 건축물의 정체성과 태도로 드러내기로 한다. 즉, 새롭게 만들어야 하는 엘리베이터, 계단실, 부속 서비스실, 연결통로, 후면 외부계단 등이 기존 건물과 어떻게 뚜렷한 의도로 구별될 수 있을까 고민했다.

모형
Model
ⓒAGIT STUDIO

'코어'부만을 집중적으로 고민한 것은 건물을 최소한의 범위로 건드려 합리적인 접근을 하는 것이 기본 전제여서지만, 동시에 '사용자'를 미리 특정할 수 없는 근생 건물에서 건축물의 '코어'는 도시와 접점을 만드는 부분으로서 사람들을 이끌고 접촉을 확장하게 하는 중요한 역할이라고 여겼기 때문이다. 그 부분이 곧 정체성을 만든다.

그 결과 새로운 기능을 담당하는 '붉은 덩어리'가 기존 건축물에 강하게 관입하듯 만들어졌다. 1단계에서 건축물 동측 부분을 철거하여 지하층 기초부터 삽입된 이 붉은 콘크리트 코어는 증축된 5층의 구조적 하중을 대부분 담당한다. 이로써 아래에 있던 기존 네 개 층은 추가되는 연직하중의 부담을 덜게 된다. 동시에 이 덩어리는 2단계인 건축물 연결 및 증축의 구조적 바탕이기도 하다. 추가 설치되는 철골 빔은 여기에 지점을 두어 형성된다.

붉은색은 의도된 덩어리를 더 강조한다. 기존 콘크리트 골조에 신설 콘크리트를 더하는 것은 동일한 습식 방식이기 때문에 구조적 강도와 물리적 일체감을 확보하지만, 오히려 변화 전후의 시간과 행위를 구분하고자 했던 의도에는 맞지 않는 측면이 있다. 애초에 그것을 이해한 접근이었기 때문에 '색'은 동떨어진 선택 사항이 아니라 처음 의도부터 자연히 존재했던 계획이다. 물량이 적어 원활하게 시멘트 공급을 받을 수 없는 등 기본적인 어려움부터 안고 있는 시공사를 대신해, 붉은 콘크리트 조색을 실현하기 위해 부산 일대 모든 레미콘 공장에 직접 연락을 취했다. 각고의 노력에도 '파봐도 진짜 빨간' 덩어리는 다음을 기약해야 했지만, 대형 장비를 오염시키거나 강도에 지장이 없는 바닥 마감 부분은 현장에서 시멘트에 조색하여 진행했다. 콘크리트에 첨가제를 사용할 경우 강도 문제와 기술적 번거로움을 포함해, 결과적으로 보여주고자 하는 표현이

Therefore, we felt that such presence should not end with merely making repairs on the building.

Rather than deriving design intent or concepts from somewhere else, we focused on revealing the changes in the building itself as it undergoes transformation according to the requirements. And we decided to manifest that act of transformation as the identity and attitude of the building. In other words, we thought about how the intent of new elements such as staircase, accessory service room, connecting passageway, and rear external staircase could be clearly distinguished from the existing building.

The reason we focused only on the "core" part is because the basic premise is to take a reasonable approach by minimizing invasive work on the building. But at the same time, given that the project involves commercial buildings where the "users" cannot be specified in advance, the "core" of such architecture is the point of contact with the city. We felt that this point of contact plays an important role in attracting people and expanding contact with them. That is to say, this part creates the identity.

As a result, a "red mass" that performs a new function was created as if it were strongly intruding into the existing building. In the first stage, the eastern part of the building was demolished, and the red concrete core was inserted from the basement foundation level, which will carry most of the structural load of the 5th floor expansion. This relieves the burden of additional vertical load on the existing four floors below. At the same time, this mass serves as the structural basis for the second stage, which is building connection and expansion. The steel beam that is additionally installed is formed by putting a supporting point here. The red color further emphasizes the intended mass. Adding new concrete to the existing concrete frame secures structural strength and physical unity because the same wet process is used, while this does not fit the intent of distinguishing between the time and action before and after the transformation.

From the start, our approach was based on such understanding. Therefore, "color" was not an isolated option for consideration, but a plan that existed naturally from the initial design intent. The company performing the construction work was having trouble in procuring cement because of short supply. So, on behalf of the construction company, we directly contacted all ready-mixed concrete factories in the Busan area to carry out the plan to mix the color red into the concrete. Despite all our hard work, we had to take a rain check on creating a red mass that "maintains its red color

같다면 굳이 그럴 필요가 있느냐라는 질문도 있었다. 나중에 컬러콘크리트 후처리 기술에 대한 긍정적인 면에 대해서도 더 알게 되었지만, 비단 비용 문제를 떠나서도 (떠날 수 없겠지만) 표면에 색을 입히는 것이 아니라 콘크리트 자체를 붉게 만들고자 했던 그 의지는 내가 생각하는 덩어리의 속성과 관계가 깊다고 생각하기 때문에 지금도 상기하는 아쉬움이다.

프로젝트 재해석
Project Re-interpret
ⓒKyungsub Shin

<프로젝트 재해석>이 보여주는 것은 완성도나 완결성이 아닌 덩어리가 가지는 메시지이다. 덩어리. 매스에 대한 강렬한 경험은 첨예한 도시문제를 만났을 때 가장 적은 비용으로 의미 있는 메시지를 만들 수 있는 기본적인 구축 어휘이기 때문이다. 우리에게 '덩어리'란 건축가가 가장 쉽게 접근할 수 있는 강직하고 기본적인 도구이다. 'Mass(매스)'가 담지 못하는 어떤 감각을 '덩어리'가 포함하는 것 같기에 이 용어를

even if it is dug out from the ground." For the floor work, we added color in the cement at the site as there was no issues such as contamination of large equipment or weakening of concrete strength. Strength issues and technical inconvenience aside, some people questioned whether mixing color additives in the cement was necessary, given that there are other methods to arrive at the desired result. Later, I learned more about the positive aspects of concrete coloring technology. In any case, aside from cost (an issue not to be ignored, though) the intent to make concrete itself red and not just applying color on the surface, is closely related to my take on the attribute of the mass. On that note, I feel regret.

프로젝트
재해석
Project
Re-interpret
ⓒKyungsub
Shin

What *Project Re-interpret* shows is not perfection or completeness, but the message intrinsic to the mass. It is possible to experience the mass intensely because it is a basic construction vocabulary that can create meaningful messages at the lowest cost when faced with acute urban problems. To architects, "mass" is an unwavering and basic tool accessible to them. We chose to use the Korean word for mass, "deongeori" because it seems to

Agit Studio

선택했다는 측면에서, 덩어리에는 우리의 미학적 태도 또한 포함되어 있을 것이다. 하지만 먼저, 재료가 무엇이건, 비싸건, 싸건 덩어리는 건축이 다룰 수 있는 가장 기본적인 제스처이자 건물이기에 가질 수 있는 고유한 속성이다. 그렇기 때문에 우리에게는 다양한 환경과 상황에서 의도를 구현할 수 있는 가장 기본적인 태도가 된다. 그리고 '모든 것은 매스 단계에서 끝남'이라는 논의가 있을 만큼 아지트 스튜디오가 말하는 덩어리는 중요하다.

프로젝트
재해석
Project
Re-interpret
ⓒKyungsub Shin

contain a certain sense that cannot be fully expressed when the English word is used. In that sense, the mass may also include our aesthetic attitude. But no matter what the material is, whether it is expensive or cheap, mass is the most basic feature that architecture can address and the unique property that a building can have. Therefore, for us, it is the most basic attitude that allows us to materialize our intent in various environments and situations. And the mass or deongeori that Agit Studio talks about is important enough to stimulate discussions on "everything complete at the mass stage."

프로젝트
재해석
Project
Re-interpret
ⓒKyungsub Shin

고집

나는 건축가들의 초기작을 보는 것에 관심이 있다. 어떤 건축가라도 그들의 젊은 날의 작업은 예산 부족, 명성 부족, 경험 부족에 따른 어려움의 집합체였을텐데, 그 과정에서도 지켜지고 남겨진 것이라면 그에게 가장 중요했던 생각과 의도일 것이기 때문이다. 내 심리가 투영되어 과하다고 생각할지는 몰라도, 가혹한 토양에서 간신히 지켜낸 어떤 것처럼 초기작에는 당시를 지배했던 생각들이 한층 뚜렷하게 드러난다.

2021년 1년간 살았던 스위스 티치노주의 리바 산 비탈레(Riva San Vitale)에는 마리오 보타(Mario Botta)의 초창기 작업인 주택 <카사 비앙키(Casa Bianchi)>가 있다. 산책코스이기도 해서 자주 들러보던 이 주택은 마리오 보타의 초창기 건축적 특성이 드러난다. 자연 위에 기하학을 만들고, 붉은색 브릿지를 이용해 경사를 극복하는 방식은 지금 우리가 아는 마리오 보타의 건축과 차이가 있지만 자신의 어휘를 노년의 대가가 될 때까지 끊임없이 발전·확장시키고 뛰어넘는 것을 그의 작품을 통해 알아보고 느낄 수 있다는 것은 내게 즐거움이자 동기부여가 되기도 했다.

아지트 스튜디오를 작업실로 운영할 때 주택 프로젝트들을 진행했었는데, 그때 인연 맺은 한 클라이언트의 소개로 <콘크리트 도서관>의 의뢰인분들을 만나게 되었다. 정식 사무실 개소 후 첫 프로젝트였다. 지금에서 돌이켜보면 의뢰인 선생님들이 처음 막연히 하셨던 이야기는, 작고 오래된 주택을 새로 쓸 수 있기 위한 인테리어 내지는 외부 모습을 좀 바꾸고 보강을 하면 어떨까 정도였던 것 같다. 하지만 그 '작고 오래된' 주택 리노베이션 <콘크리트 도서관>은 그 후 2년이 넘는 시간을 함께하는 프로젝트가 되었.

길이 30m의 좁은 골목길 끝에 30년 연식의 노후 연와조 주택이 앉아 있고, 폭이 2.5m도 채 되지 않는 좁은 골목길은 '대지'임에도 방치된 골목처럼 놓여 있다.
상황1_ 철거 후 신축이 불가, 즉 사실상 맹지, 상황2_ 주차확보 불가능, 즉 증축 불가능, 상황3_ 기존 구조체 노후, 지반침하 진행 중.

전, 후 모습
Before, after
ⓒAGIT STUDIO
ⓒKyungsub Shin

Persistence

I am interested in seeing the early works of architects. For any architect, the work they did early in their careers is probably a collection of difficulties due to the lack of budget, reputation, and experience. If the architect preserved and maintained those works in the midst of those difficult processes, it would be because those works incorporated the thought and intent considered most important by the architect. Although this statement may sound excessive reflecting my conjecture, the early works of an architect clearly reveal the thoughts dominating the architect at the time, as if a delicate being was barely preserved in harsh soil.

I lived in Riva San Vitale, Ticino, Switzerland for a year in 2021. There, you can find *Casa Bianchi*, a house designed by Mario Botta early in his career. As *Casa Bianchi* is located along a trail, I often visited the house because it reveals the architectural characteristics of Mario Botta early in his career. The house was built in nature using geometric forms and a red bridge was applied to overcome the limitation of the slope. Such methods are different from the architecture of Mario Botta that we know today. However, it was a pleasure and motivating for me to be able to recognize and feel his vocabulary through his works even as they constantly developed, expanded, and surpassed until he became a master at an older age.

When I was running Agit Studio as a workshop, I worked on house projects. One of the clients of those projects introduced me to a couple who would later be the clients of the *Concrete Library* project, which is the first project I worked on after opening Agit Studio officially. Looking back, the first thing the clients told me was that they wanted to do some interior work or change and strengthen the exterior of a small, old house so that it could be reused. However, the renovation of that "small, old" house became a project called *Concrete Library* that took over two years to complete.

A 30-year-old house sits at the end of a 30-meter-long narrow alley. Less than 2.5 meters wide, the narrow alley looks like an abandoned alley even though it is categorized as a "site." Three odds apply here. First, the house cannot be rebuilt after demolition, meaning its site is effectively a landlocked land. Second, it is impossible to secure parking space, which means expansion work is impossible. Third, the existing structure is deteriorating, and ground settling is in progress.

Agit Studio

제대로 할라치면 총체적 난국인 부지에서 해결 방안에 접근하는 방식부터 의견을 좁히는 게 쉽지 않았다. 허근일은 예산과 공사 기간 등 합리적인 이유를 들어 비록 우리가 좋아하는 방식은 아니지만 필요한 부분에 철골보강을 하고 외장타일을 변경하는 등의 현실적이고 손에 잡히는 범주부터 접근해야 한다고 말했다. 하지만 나는 완전히 다른 생각이었다. 임시 해결이 아닌 근본적인 방법을 찾는 것에서부터 시작하기를 주장했다. 더욱이, 의뢰인 선생님들은 이 집을 평생의 '마지막 집'으로 생각한다고 하지 않았나. 무엇보다 나는 하중을 견디지도 못할 슬래브와 기초에 철골 기둥을 지점별로 세워 보강하는 건식 방식에 불신이 있었다. 근본적인 접근과 건물의 영속성을 강조한 방식은 프로젝트의 시작점에서 중요한 의도가 될 수 있었고, 이후 고민의 초점은 지반침하가 일어나는 상황에서 구축적 해결방식을 찾는 것으로 바뀌었다. 이 작은 규모의 건물에서 법적, 행정적, 기술적, 비용적 어려움에 맞서보겠다는 이야기였다. 유의미한 건축 계획이 되어야 하는 것은 물론이었다.

기존 벽돌벽과 신설 콘크리트 벽
Old brick wall and new concrete wall
ⓒAGIT STUDIO

<콘크리트 도서관>은 오랜 계획 기간과 지난한 공사과정을 감당해야 했다. 기존 벽돌 벽체를 콘크리트의 '폼(거푸집)'으로 사용하자는 아이디어는 긴 계획 협의와 디테일 연구를 거쳐 마침내 실시 도면으로 완성되었다. 그러나 건축물 부분 철거가 시작된 후 드러난 단배근의 슬래브, 흙바닥 위에 그저 놓여있는 부실한 기초, 벽체마다 발생한 크랙 등과 마주했을 때는 건축가도, 시공사도 당황할 수밖에 없었다. 그리고 다시 시작된 설계변경 과정을 거쳐 60평 남짓의 건축물을 완성하기 시작했다. 그 과정은 집요함이 만든 완성이기도 했다. 프로젝트를 적극적으로 이끌어야 하는 나보다는 프로젝트를 객관적으로 볼 수 있는 입장이었던 허근일이 말했다. 이 작업의 의미는 계획 초기에 철골 보강 건식 방식이 아니라 영구적 습식 방식으로 가고자 했던 의지를 보였을 때, 그리고 그 방향으로 진행 합의를 이루었을 때 이미 정해진 것이었다고.

In such a poor site condition, it was difficult for us to do proper work there and even to narrow the gap between differing opinions on the approach to solve this problem. Citing rational reasons such as budget and construction period, Guenil Huh said that we should have an approach which starts from realistic and manageable categories such as steel frame reinforcement and exterior tile replacement where necessary, albeit it is not our preferred approach. But my idea was totally different from his. I insisted that our work should start from finding a fundamental solution rather than opting for a temporary solution. What is more, the clients considered this house to be the "house they want to live out the remainder of their lives in." Above all, I had distrust in the dry method of reinforcement such as using slabs and installing steel columns at different points, which I felt would not be able to withstand the load. At the start of the project, it was an important intent to take a fundamental approach and emphasize the building's permanence. Later on, the focus of our consideration shifted to finding a structural solution in situations where ground subsidence occurs. The idea was to overcome legal, administrative, technical, and cost difficulties in connection with this small-scale building. Of course, the architectural plan had to be meaningful.

To complete *Concrete Library*, we had to endure a long planning period and an arduous construction process. The idea of using the existing brick wall as the concrete's "form" finally materialized in the working drawing after lengthy planning consultations and extensive detail studies. However, after partial demolition of the building began, both the architect and the contractor could not help but be taken aback when encountering the exposed single reinforcement slab, the weak foundation placed on a soil floor, and the cracks that appeared on every wall. With that, the design change process started, and the process to complete the building of about 60 pyeong (approximately 198 square meters) progressed. That process is also the completeness borne from persistence. Unlike me, who was actively leading the project, Guenil Huh was in a position to view the project more objectively. He said that the meaning of this work had already been decided at the beginning of planning when the intent to go with a permanent wet construction rather than a dry construction of steel reinforcement was expressed, and subsequently when it was agreed upon to proceed in that direction.

콘크리트 도서관
Concrete Library
ⓒKyungsub Shin

<콘크리트 도서관>에서 지켜내고 남긴 것이 무엇인지 스스로에게 되물어본다. 그에 대한 답은 꽤 명료하다. 그 명료함이 과연 좋은 것인지 아닌지는 판단하기 어렵지만 말이다. 이 작업에서 가장 중요하게 정리하고 다룬 의도 - 기존 연와조 구체를 활용한 콘크리트 구축 - 는 다른 부분들을 모두 폐어 냈다. 온갖 번잡한 범위를 아우르고 해결해야 하는 건축계획의 과정에서 날카롭게 수렴하는 지점을 만들어 폐어 나간다는 것은 고도의 지적 작용이다. 이것은 나에게 가장 중요하고, 큰 괴로움과 동시에 즐거움을 주는 창작과정이라고 말할 수 있다.

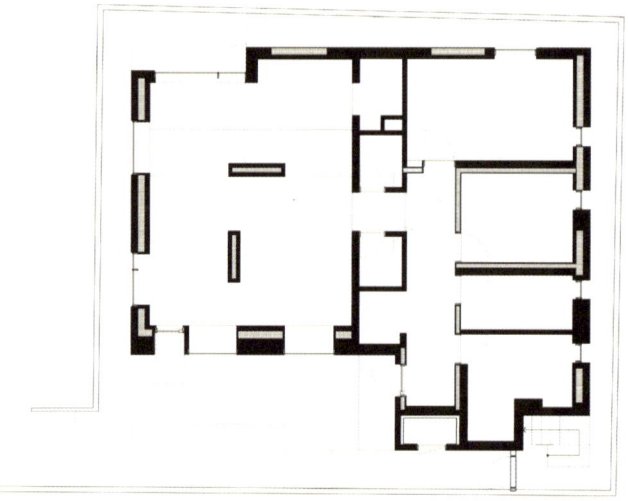

1층 평면도
1F Floor plan
ⓒAGIT STUDIO

I ask myself what was preserved and kept with *Concrete Library*. The answer is pretty clear, but it is difficult to determine whether that clarity is a good thing or not. However, the most important intent in this work, concrete construction which uses the existing brick frame, pierces through all other parts of the project. The architectural planning process encompasses and resolves various complex categories. Creating sharp convergence points and stitching them together involves sophisticated intellectual work. This is a creative process that brings me great pain and joy at the same time.

기존 30년의 흔적은 그대로 유지된 채로 새로운 레이어가 덧입혀진다. 시멘트 조적벽과 콘크리트 벽은 합벽이 되어 60㎝ 이상의 두꺼운 두께가 되고, 이는 도서관 내부에서 깊이를 만든다. 깊숙이 들어온 빛이 작은 공간에 볼륨과 리듬을 만들어낸다. 단단하게 비워진 공간을 지탱하기 위해 있는 중앙 벽체도 기존 조적벽에 콘크리트가 보강된 것이다. 문제 해결을 위한 구축 방식은 1층 도서관에 정체성을 부여하고 있다. 반면, 기존 조적벽과 신설 콘크리트 벽이 거리를 두어 생긴 공간은 새로운 테라스가 된다. 붉은 타일이 붙어있던 옛 조적벽에는 빨간색 스토가 입혀져서 지난 시간을 드러내고 있다. '도서관'이라는 프로그램은 평생 책과 연구를 가까이 해온 의뢰인분들의 '책'과 관계된 공간을 만들고 싶은 바람에서 출발한 것인데, 2층 주택에 속해 작은 부분으로 존재할 수도 있었지만, 지속적인 계획과정과 미팅들을 통해 건축물의 중심 프로그램으로 자리잡게 되었다. 방치되었던 골목길은 이제, 사람들의 발걸음을 세워 무심코 걸어 들어오게 만든다. 비워낸 콘크리트 덩어리는 누구나 방문할 수 있는 도서관이 되었다.

콘크리트 도서관
Concrete Library
ⓒKyungsub Shin

우리는 <콘크리트 도서관>을 '콘크리트 텍토닉'이라고도 부른다. 이 말은 당연하게 여겨지는 방식을 벗어나 우리가 고민한 구축 방식을 적용하고 모든 공정, 내러티브, 공간, 디테일 등을 콘크리트로 일관되게 풀어내어 건축물의 모든 부분을 다루었다는 것을 뜻한다. 콘크리트는 구조적 문제 해결의 주체로서, 동시에 공간을 비우고, 깊이를 만들고, 장소를 정의한다. 그 재료적 물성은 내외부 재료로도 십분 활용된다. 위치에 따라 거칠게, 더 거칠게, 부드럽게, 혹은 날카롭게 가공되어 의도에 맞는 질감을 형성한다.

콘크리트를 이용한 이러한 구축방식은 이후 작업한 <프로젝트 재해석>에도 영향을 끼쳤다. 하지만 무엇보다, <콘크리트 도서관>을 통해 남긴 것은, 의도를 예민하고도 실질적으로 만들어 프로젝트의 큰 부분에서 작은 부분까지 관통하고자 하는 방식에 대한 노력의 결과였고, 그것이 좋은 건축을 만들 수 있다는 우리의 유의미한 고집이었다.

A new layer was added while traces of the previous 30 years were maintained. The cement masonry wall and concrete wall became a thick composite wall of more than 60 cm. This has created depth inside the library. Light enters deep into the small space and creates volume and rhythm there. The central wall that firmly supports the empty space is also the existing masonry wall reinforced with concrete. The construction method used for problem solving gave the library on the first floor its identity. This so-called "library" program started from the desire of the clients to create a space related to books as they have spent most of their lives reading books and doing research. It might have remained a small part of a two-story house project, but through the planning process and several meetings, it eventually became the central program of the building. On the other hand, the space between the existing masonry wall and the new concrete wall was planned as a new terrace. The old masonry walls which had red tiles were rendered with red Sto finishing, revealing the past. And the alley, which had been neglected, now draws people to it. In this way, the emptied concrete mass became a library that anyone can visit.

콘크리트
도서관
Concrete
Library
ⓒKyungsub
Shin

We also refer to *Concrete Library* as "Concrete Tectonic." This means that we did not opt for the reasonable method of architecture but applied a method derived from deep consideration of all parts of the building including all processes, narratives, spaces, and details with this intent in mind. Therefore, the words contain our thought and expectation that these feelings are conveyed. Concrete began as an agent of structural problem solving, while simultaneously emptying space, creating depth, and defining the place. Properties of this material are such that can be fully utilized as interior and exterior materials. Depending on the location, it is processed coarsely, roughly, softly, or sharply to create a texture that suits the intent.

This tectonic method of using concrete also had an influence on the subsequent project, *Project Re-interpret*. However, *Concrete Library* was the result of our efforts to make our intent sensitive and practical, making it pierce through the project from the large to the small parts. It also reflected our meaningful persistence grounded in the belief that such efforts can create good architecture.

콘크리트 도서관
Concrete Library
©Kyungsub Shin

콘크리트
도서관
Concrete
Library
ⓒKyungsub
Shin

태도 : Attitude

Agit Studio

못생김

못생김에 대한 논의는 용기가 필요할지도 모른다. 이는 우리가 가진 건축적 미학에 관한 태도이다. 우리 작업은 절대적인 미학적 기준이나 비례를 설정하거나 추구하지 않는다. 건축물이 배경처럼 존재해야 할 때와 존재감을 드러내야 할 때는 각 프로젝트와 대상지마다 다르다. 우리가 계획을 진행하는 방향이 끊임없이 의도(주제)를 명료하게 하기 위한 과정이듯, 그 과정에서 형태를 만들고 매스를 논의할 때도 도형·조형적으로 불필요함 없이 명쾌해지는 방향으로 지속적으로 발전시킨다.

우리가 계획과정에서 만나는 현장은 시대를 걸쳐 건축가들이 고심해 만들어놓은 도시 위에 있지 않다. 도시 인프라 구축이라는 필요성에 의해 급하게 찍어내듯 만들어졌거나, 어쩌면 경제 논리에 의해 과장된 형상을 지닌 도시에서 이루어진다고 말할 수도 있겠다. 하지만 유럽의 도시들처럼 치밀한 마스터플랜 위에 건물을 하나하나 만들어낸 정성은 없을지라도, 나는 불규칙하고 역동적인 우리 도시에서도 미학을 찾을 수 있다고 생각한다. 이것이 '못생김'을 우리의 태도 중 하나로 말하려는 이유이다.

아카이브 매스
Archived Mass
ⓒJoel Moritz

Ugliness

Discussing ugliness may require courage. This is our attitude toward architectural aesthetics. Our work does not set or pursue absolute aesthetic standards or proportions. Situations where a building should exist as a background and where it should reveal its presence are different for each project and site. The direction of our planning is a process of constantly clarifying the intent (subject). Likewise, when creating form and discussing the mass in that process, our plan continues to develop toward a clearer direction without figurative and formative redundancy.

The site we encounter in the planning process does not exist in the city that architects have, for a long time, taken great pains to create. It may have been created in a hasty manner as when something is mass-produced in a factory due to the need to build urban infrastructure, or, created in the city with exaggerated shapes due to economic logic. However, even though there may not be the same level of care in creating each building based on a detailed master plan as in European cities, I believe that aesthetics can be found even in our irregular and dynamic cities. This is why I want to talk about "ugliness" as one of our attitudes.

<아카이브 매스>는 미술품을 전시하는 갤러리인 만큼 하나의 순수한 화이트 큐브가 될 수도 있었다. 실제로 작은 근생 주택을 갤러리로 바꾸고 싶다는 의뢰가 들어왔을 때 첫 제안으로 생각한 것은 건축물 전면과 주변을 금속을 이용하여 깨끗하고 세련된 선으로 정리한 모습이었다. 그것은 골목에서 오브제로서 꽤 아름답고 아이코닉한 갤러리가 될 수 있다는 가능성과 남다른 형상을 만들고자 하는 우리의 욕심을 담고 있었다. 이러한 계획안을 가지고 연희동 동네를 다시 가보았다. 그곳은 정감이 남아있는 시장길 동네 같기도 했다. 아직 개발의 바람이 닿지 않은 작은 주택들과 3, 4층의 근생 건물들로 여전히 동네라는 모습을 형성하고 있는 대상지는 역시 가장 보편적이고 익숙한 '제2종 일반주거지역'에 있었다. 흥미롭게도 부지가 있는 거리는 하나둘씩 뜨문뜨문 작은 갤러리 전시공간으로 변화하고 있었다. 이 장소에서 모습을 바꾸어 새로 태어날 갤러리는 어떤 태도를 보여야 할 것인가? 이 질문은 분명 아름답거나 특별하거나 잘생긴 것을 만들어야 한다는 것과는 다른 입장에서 <아카이브 매스> 작업을 진행하게 했다.

연희동 골목에서 마주한 건물은 특별한 이력을 가지고 있었다. 태생부터 주인이 달랐던 두 필지에 똑같은 모습의 3층짜리 쌍둥이 건물이 서 있었는데, 서로 지하층 출입구를 공유하고 있었다. 땅 위로 데칼코마니 같은 건물 두 개가 지하에서 묶여있는 상태였던 것이다. 한쪽 건물만 건드릴 수 있는 상황이었지만 우리는 이 세트 건물의 스토리 라인을 이어가고 싶었다. 그래서 가로에서도 중요한 입장이자 쌍둥이 건물에서 가장 주요한 특징인 '입면'을 존중하기로 한다.

사실 이번 작업과 관계없는 옆 세탁소 건물을 같이 두고 본다. '갤러리'와 세탁소는 나란하면서도 독립적인 모습을 갖추어야 할 것이다. 본래 건물을 구성하는 요소들을 각각 분리시켜 생각해 보았다. 기존 입면을 존중하겠다는 것은 창문과 개구부를 곧이곧대로 유지하겠다는 말은 아니다. 새로운 계획에 따라 창이 다시 구성되지만 기존 개구부 흔적을 유지한다. 스토 마감의 스크래퍼를 다르게 사용하여 입자를 달리하는 방식은 새 역할에 맞게 창문은 작아지더라도 본래 있던 창의 자리가 이질적인 질감을 가지고 남아서 드러나도록 하였다.

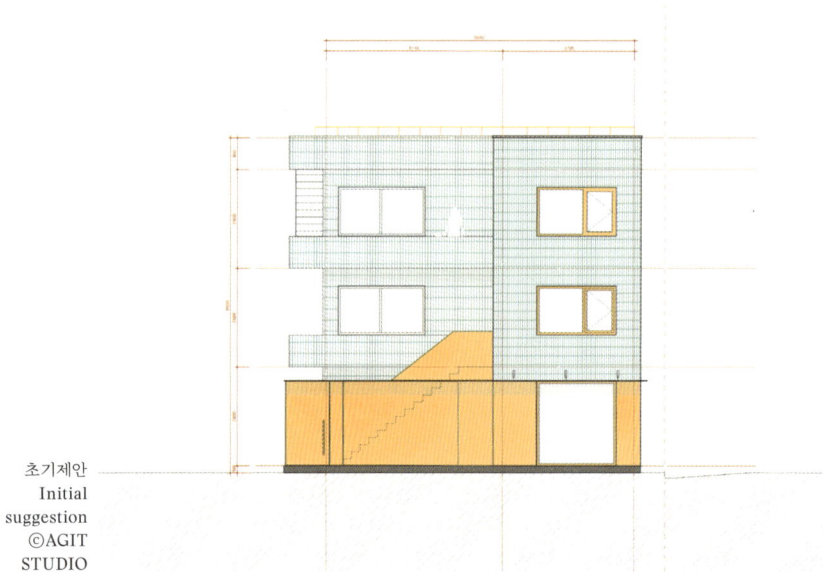

초기제안
Initial suggestion
ⓒAGIT STUDIO

As *Archived Mass* is a gallery that displays works of art, it could have been a pure white cube. In fact, when we received a request to turn a small retail and single-family house into a gallery, the first thing that came to mind was to propose a neatly organized building with clean and sophisticated metal lines on its front and its surroundings. It contained both the possibility of becoming quite a beautiful and iconic gallery as an object in the alley and our desire to create a unique shape. With this plan in mind, I went back to the neighborhood of Yeonhui-dong. The place felt like a market street neighborhood with lingering warmth. It was an area which still maintains the appearance of a neighborhood with small houses and 3- and 4-story retail buildings and houses, not yet touched by the wind of development. The project site is in "class II general residential area," the most common and familiar area category. Interestingly, buildings here and there along the street where the site is located were turning into small galleries or exhibition spaces. What attitude should the gallery that will be reborn in this location exhibit? This question led me to work on *Archived Mass* from a different perspective, not driven by the need to create something that is clearly beautiful, special, or good-looking.

The building we encountered in the alley of Yeonhui-dong had a special history. Two identical three-story buildings stood on two lots that had different owners from the beginning, and they shared a basement entrance. In other words, the twin buildings were joined at the basement. Although we were commissioned to work on only one of the twin buildings, we wanted the story line of the two buildings to continue. Therefore, we decided to respect the "facade," which is an important street element and the most important feature of twin buildings.

In fact, my approach was to consider the other building which had a laundromat on the first floor, even though the building had nothing to do with the project. The gallery and the laundromat should look as if they are arranged side by side but independent as well. Original elements that make up each of the twin buildings were considered separately. Respecting the existing facade does not mean that the windows and openings will remain exactly as they are. The windows were reconfigured according to the new plan, but traces of the original openings were retained. Thus, even though the size of the windows has become smaller to suit their new role, using different scrapers in the Sto finishing to vary the particles allows the location of the original windows to be retained with the heterogeneous texture.

Gypsum wall boards were newly installed for the main gallery space on the first floor. These clean boards were placed in juxtaposition with the existing crooked concrete beams on top, right beneath the ceiling.

아카이브 매스 Archived Mass ©Joel Moritz

1층 메인 갤러리 공간을 위해 다시 깨끗하게 설치된 석고보드 벽체는 상부의 삐뚤빼뚤한 기존 콘크리트 보와 병치된 채로 놓았다. 전시 관람객이 이것을 보고 처음에는 마감이 덜 되었나 생각할 수도 있다. 누군가는 지금이라도 실런트를 채워 넣는 게 어떻겠냐고 말하기도 할 것이다. 그렇지만 곧 울퉁불퉁하게 불협한 라인을 보면서 이 갤러리는 이전부터 있던 장소를 바탕으로 했음을 상기하면 좋겠다. 못난이 보처럼 멋스러움을 느낀다면 더욱 좋을 것 같다. 하지만 전시물을 설치할 때 수평 기준을 잡는 데에는 좀 노력이 더 든다는 피드백은 기억에 남겨두었다. 몇 차례 리노베이션 작업을 하면서 반복적으로 느끼는 것이지만, 우리네 기존 건축물들은 그렇게 '못생기게' 남아서 도통 정확한 기준 지점을 찾기 어렵게 만든다. 지하층 갤러리에서는 곰팡이 슨 벽지나 떠붙이기로 시공된 타일을 걷어낸 후 그래도 보기에 지저분한 부분들은 일부 벽면을 갈아내었는데, 톱날 흔적이 마치 묵혔던 때를 벗겨낸 모습으로 그림처럼 나타났다. 외부마당에서 접근한 방식도 동일하다. 무작위로 덧댄 슬레이트 지붕을 받치는 엉터리 벽 역할을 하고 있던 시멘트 담장을 본 모습으로 되돌리고, 그것을 이용해 마당을 즐길 수 있는 소소하고 독특한 콘크리트 벤치와 화단을 계획했다. 모두 아름답지 않았던 기존의 것을 재해석하여 의미있는 것으로 만드는 과정에서 생겨난 것들이다.

아카이브 매스
Archived Mass
©AGIT STUDIO

훈련된 감각을 이용하여 아름다움을 만드는 것은 건축물로서, 또 건축가에게 매우 큰 의미이다. "Less is more"라는 근대 담론을 넘어가는 듯한 지금의 건축세계는 더욱 미학에 새로운 태도를 보이는 것 같다. 그러나 우리가 도시에서 직접 마주하고 다루는 환경은 단독으로 아름답기에 좀 곤란할 때가 많다. 어려운 것은, 그것을 불가피하게 안고 가야만 할 때이다. 누가 봐도 남겨둘 가치가 있는 아름다움을 유지하는 것과는 다르다. 30년 연식의 찍어낸 건물들을 존중하는 바탕 위에서 우리만의 해석을 하는 것은 매번 당면하는 일이다. 그 존중이란 건 감상이 아닌, 보통은 철저한 현실의 여건을 바탕으로 한다.

<아카이브 매스>는 숨어있는 건물의 원형을 찾아내 그것만 드러냈다. 그 원형은 사실 없었던 것일 수도 있다. 그것은 내재된 본래의 DNA일 수도 있고, 실은 간신히 건져냈거나 새로 주입한 것일 수도 있다. 중요한 것은 '못생긴' 원형을 다루는 것에서 시작했던 우리의 관심과 생각이다. 아마도 못생김은 해학적이고 반항적인 심사를 다룬 말일 것이다.

When visitors to an exhibition see this, they may initially think that the construction work has not been finished. Some people may even suggest that those gaps be filled up with sealant. However, I wanted people to see those uneven lines and be reminded that the gallery is based on the place that had always existed there. It would be even better if they could feel a sense of style from the ugly beams. However, I did take note of the feedback that it takes more effort to achieve horizontal alignment when the exhibits are installed. There is something that I feel repeatedly every time I do renovation work. Because what remains of the existing building is "ugly," it is difficult to find an accurate point of reference. In the basement gallery, the moldy wallpaper and glued tiles were removed. Parts of the wall that still looked messy were ground off with a saw blade. It turns out that traces of the grinding looked like a picture depicting how the old stain was removed. The approach to the external yard is the same. The crude cement fence that supported the randomly added slate roof was returned to its original state, so as to make a small and unique cement bench and a flower bed where one can enjoy the yard. All this was derived from the process of re-interpreting the existing things that are not beautiful into the things that are meaningful.

아카이브 매스
Archived Mass
ⓒJoel Moritz

Using trained senses to create beauty holds great significance to architecture and the architect alike. The current architectural world, which is moving beyond the modern discourse of "Less is more," seems to display a new attitude toward aesthetics. However, when we face or deal with the environment in the city, it is often difficult to jump to the conclusion that it is beautiful in its own right. This is especially the case when you have no choice but to embrace it. That is, this case is different from maintaining beauty that is deemed to be worth preserving to anyone's eyes. This is something we face every time we do this sort of project – overlaying our own interpretation based on respect for 30-year-old buildings which look to be printed out of factories. Here, respect is not based on sentiment, but is usually based on outright realistic conditions.

In *Archive Mass*, the hidden original form of the building was found and only that part was revealed. In fact, the original form may not have actually existed in the first place. It may be the original DNA embedded in the building, or something that may have been barely salvaged or newly injected. Whatever the case, what is important is our interest and thoughts that started from dealing with the "ugly" original form. Perhaps ugliness is a word that addresses humorous and rebellious state of mind.

아카이브 매스
Archived Mass
ⓒJoel Moritz

Attitude of AGIT STUDIO | 태도: Attitude

아카이브 매스
Archived Mass
ⓒJoel Moritz

Agit Studio

내러티브

2020년 여름, 각 세대에 폭이 깊은 외부 테라스가 있는 다세대주택 <모따기99>가 완공되었다. 계획과정 동안 즐겁게 호흡이 잘 맞았던 건축주는 모든 실이 계약 완료되었다는 사실을 알려주었다. 여러 해를 걸치기 마련인 프로젝트들이 마침내 완공될 때 느껴지는 감정들에 더해서, 또 중요한 이 소식은 예상치 못한 해방감을 주는 듯했다. 같은 '주거'라는 프로그램이지만, 임대가 중요한 공동주택 작업은 프로세스와 의미 차원에서 단독주택 계획과 큰 차이가 있었다. 아파트 개발예정지로 묶여있다가 계획이 풀린 일반주거지역에 지어진 공동주택 프로젝트였다.

이후 누군가 <모따기99>를 두고 코로나 이후에 테라스 있는 집이 잘 나간다던데 그래서 이렇게 설계한 것이냐고 물어오기도 했지만, <모따기99>는 코로나 바이러스가 세상을 지배하기 이전에 계획을 마치고 착공에 들어간 작업이었다. 그 질문을 받는 순간 나는 전염병마저 곧 유행으로 치환되어 팔리는 현상을 직면한 것 같았다. 역병이든, 사고든 일순간에 우리 사회를 지배하는 '유행'보다 우선하는 '내러티브'가 건축에 존재한다. 그리고 그것은 우리가 작업을 이어온 토대 중 하나다.

<모따기99>는 그 모양새가 흥미를 끄는 것 같다. 조형성이 특징으로 보이는 이 프로젝트는 단순한 도형을 다각적으로 반복 이용하면서 의도한 덩어리를 만들어낸 것이지만, 어떤 작업보다도 오히려 형태적인 정의에 있어서는 여러 질문을 바탕으로 진행되었던 계획의 결론적인 단계에서 정의되었던 것 같다.

모따기99
Mottagi99
ⓒKyungsub Shin

Narrative

In the summer of 2020, *Mottagi99*, a multi-family house with each unit having a deep external terrace, was completed. Our team and the client got along very well during the planning process, and the client later informed us that rental contracts for all units have been concluded. One feels a unique sense of emotion when a project, which extends several years, is finally completed. Furthermore, when the client gave us the good news, I felt liberated, an emotion which I hadn't expected to feel. The project falls under a "housing" program, but in a multi-family home project, rental is an important factor. In that sense, this project is very different from a single-family home project in terms of process and meaning. *Mottagi99* is a collective housing project built in a general residential area, where new building development had been regulated because of the planned development of an apartment complex but later was deregulated.

Someone asked me whether *Mottagi99* was designed as such because of the trend after the COVID-19 outbreak of increasing popularity of housing units with terraces. In fact, project planning completion and groundbreaking for Mottagi 99 took place before the coronavirus pandemic took the world by storm. So, the moment I was asked that question, I felt like I was witnessing a situation where even infectious diseases were being transformed into a marketing trend. However, in architecture, there is the existence of "narrative" which takes precedence over "trends" that dominate our society at that moment whether it is a pandemic or an accident. And it is also one of the foundations that enables our work to continue.

First, *Mottagi99* is interesting in its shape. This project, which is characterized by formativeness, created the intended mass by repeatedly using simple shapes in various ways, but the definition of the building's form was determined at the concluding stage of project planning, which was based on a multitude of questions.

Agit Studio

우리가 접하는 공동주택들은 정형화된 형식과 논리로 존재한다. 아파트는 '이렇게' 생겼고, 빌라라고 불리는 다세대·다가구 건물은 '이렇게' 생겼다. '이렇게' 생긴 모양은 아마 부동산 개발 주체에 의해 한 치의 땅 낭비 없이 작은 수익까지도 계산되어 만들어진 가장 합리적인 요지부동 정답으로 여겨진다. 우리는 다세대 프로젝트를 하면서 규제와 경제성의 산물로 탄생한 이 타이폴로지(유형)에 감히 도전하고 싶어졌다. 2016년 베니스 건축비엔날레 한국관의 '용적률 게임'이라는 주제처럼 우리만의 방식으로 이 문제에 대한 다른 답을 구해 보고 싶었다. 절대 답 이외에 없을 것 같은 영역에 창의적인 이야기 하나를 더할 수 있다면 그것으로도 괜찮은 시도일 것이었다.

우리가 뽑아낸 다세대주택의 '전형'이라는 것은 이렇다. 지상 1층은 전부 필로티 주차장으로 되어있기 때문에 보행자는 언제나 양측으로 차들이 가득찬 어두운 주차장을 마주한다. 동시에 개발된 것이 아님에도 불구하고 때로는 거대한 하나로 계획된 주차장을 걷는 기분이 든다. 이 우울한 집단적 상황은 거주하는 사람들에게 필요한 동네의 맥락이나 필요한 시설과 인프라를 만들어 줄 수 없다. 도시적 관점에서의 풍경과 맥락도 안타까울 뿐이다. 또 한 가지는 일조권 사선제한선에 맞춰 결정된 형태들이다. 이 형태는 옆집의 남향 일조를 확보해 주기 위해 내 집 북측을 사선으로 깎아 내준 것이다. 상부상조 원칙처럼 보이는 이 규제는 정북 방향으로 일제히 귀퉁이가 썰려 나간 도심 주거 풍경을 만들었다. 사실 여기에 '준공 후 건축'이라는 이상하고도 당연하게 만연된 불법 증축 행위가 도심 주거지의 이질적 모습을 심화시킨다.

우리에겐 '용적률 게임'에 적극적으로 임하면서도 타이폴로지에 도전하고자 하는 과감함이 필요했다. 계획의 원칙은 당연한 이 전형들에 정면으로 반기를 드는 것이었다. 1층을 주차장으로 점유시키지 않고 동네에 필요한 시설을 담는 프로그램으로 계획하는 것, 그리고 사선제한이 형상을 지배하지 않도록 하는 것이었다.

모따기99
Mottagi99
ⓒKyungsub Shin

The collective houses that we encounter exist according to the stereotyped shape and logic. Apartments have to "look a certain way" and multi-family or multi-unit houses, otherwise referred to as villas, have to "look a certain way." The shape of "this certain way" was probably determined with confidence by the entities involved in real estate development by calculating even the smallest profits and not wasting an inch of land. While working on the multi-family project, we ventured to challenge this typology (type) that was born from regulations and economics. We wanted to find a different answer to this problem in our own way, just like the theme of the Korean Pavilion at the Venice Architecture Biennale 2016, "FAR (floor area ratio) game." If we can add one more creative story to an area where there seemed to be only one absolute answer and no other, that in itself would be deemed a good attempt.

Our take on the "stereotype" of multi-family housing existing in Korea is as follows: The first floor is pilotis and is entirely used as a parking lot. Therefore, what pedestrians see on both sides is a continuum of dark parking lots filled with cars. Even though the buildings were not built at once, it sometimes feels as if one is walking through a single large parking lot that was originally planned that way. This depressing collective situation cannot cater to the needs of the residents such as creating a context for the neighborhood and providing necessary facilities and infrastructure. The scenery and context from an urban perspective are deplorable. Another thing is that the shape of buildings is determined according to the diagonal setback regulation for daylight. This shape is the result of cutting the north side of a building diagonally to secure south-facing sunlight for the building next door. This regulation, which seems to be based on a principle of mutual assistance, created an urban residential landscape in which corners of building's side facing the north are sloped. Moreover, the strangely widespread practice of illegal expansion referred to as "post-completion construction" aggravates the heterogeneity of urban residential areas.

What we needed was the boldness to actively play the "floor area ratio game" and at the same time, to challenge typology. The principle of our plan was, of course, to revolt against these given stereotypes. That is, plan a program which accommodates facilities needed in the neighborhood on the first floor, rather than a parking lot. Also, to not let the setback regulation dominate the shape.

다시, <모따기99> 2, 3층에는 깊은 테라스를 가진 세대가 둘씩 있다. 거리에서 시선을 사로잡는, 크게 비워진 상층부 공간 역시 외부 테라스이다. 4, 5층은 의뢰인 가족이 사는 집인데, 이 큰 테라스 덕분에 빽빽한 공동세대 중 하나가 아닌, 단독주택에 사는 효과를 가질 수 있다. 주택 프로그램임에도 불구하고 우리가 시종일관 주안을 두었던 이야기는 외부의 힘에 관한 것이었다. 외부의 작용들이 덩어리를 조물하듯 영향을 끼쳤다. 실, 코어, 외부공간 등 밀도의 재배치가 진행되는 과정 속에서 계획이 다듬어졌다. 개인 주거공간의 질이나 꾸밈에 대한 것에 우선하여, 근본적인 제약을 만드는 것들에 의문을 품는 것이 주거의 기본을 좋게 할 수 있다고 생각했다.

1층 공동 출입구는 의도적으로 감추었다. 큰 아크 곡선을 그리며 무의식적으로 진입을 이끌게 하기 위함이었는데, 그 곡선 매스는 곧 소매점이 들어가는 프로그램이기도 했다. 우리는 이를 법정 주차대수, 조경·생태면적, 구조 전이부, 이격거리 등 지상층을 지배하는 여러 상관관계를 고민하며 세밀하게 조정했다. 무조건 세대수를 늘려야 이롭다는 지역부동산의 의견에 강하게 맞선 순간도 있었지만 처음에 세웠던 원칙을 건축주와 복기하며 이를 극복해 나갔다. 완공이 가까워진 어느 날 오후, 동네 아이들이 새로 생긴 건물 주변을 돌아다니며 저마다 1층에 무엇이 생기면 좋을지 떠드는 모습을 지켜보았다.

서번트 공간을 최대한 콤팩트하게 만들어 실의 면적을 최대화하겠다는 의지는 계단실에 구조미를 발현시키도 했다. 계단참을 아낄 수 있는 돌음계단은 중앙에 놓인 구조 벽체에 지지해 있는데, 이와 동시에 계단실 핸드레일의 폭까지 아끼기 위해 이 벽체에 손스침을 삽입했다. 번거로울 수 있는 공사 과정이었지만 의도를 전달하며 시공사를 설득했다.

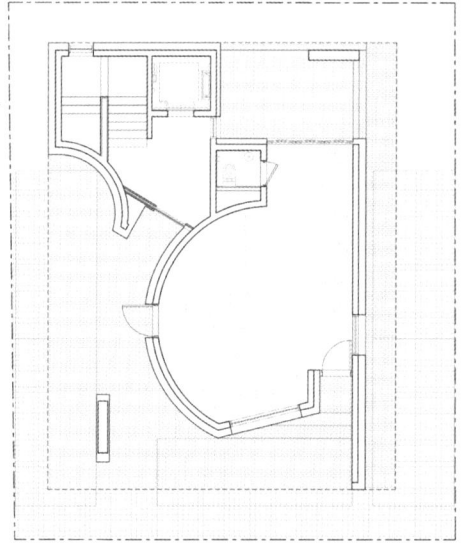

1층 평면도
1F Floor plan
ⓒAGIT STUDIO

On the 2nd and 3rd floors of *Mottagi99*, there are two units with deep terraces on each floor. The largely empty space on the upper floor that catches people's attention from the street also is an exterior terrace. The 4th and 5th floors are where the client's family lives, and thanks to this large terrace, the family can feel like they are living in a single-family house rather than as occupants of a crowded multi-family house. Even though we were working on a housing program, our main focus throughout was on external forces. The external forces at play affected the mass as if they were fumbling with it. The plan was polished in the process of rearranging the density of rooms, the core and external space. We thought that before considering the quality and decoration of personal living space, questioning the things that create fundamental constraints, would improve the basics of dwelling.

The common entrance on the first floor was intentionally hidden. The purpose was to induce people to unconsciously enter the building following the large arc-shaped curve. The curved mass also accommodates the program of retail space. We adjusted its details by carefully considering the correlations of factors that govern the ground floor, such as the legal parking space, landscaping/ecological area, structural transition area, and separation distance. There were moments when we had to take a strong stance against the local real estate agency's opinion that it would be beneficial to unconditionally increase the number of households, but we overcame this situation by revisiting the principles we had established at the beginning with the client. One afternoon, as the project was nearing completion, I watched some neighborhood children walking around the building and overheard their conversation about what store they wanted to have on the first floor.

The will to maximize the room area by making the servant space as compact as possible also manifested the structural beauty in the staircase. The winding stair that saves landing space is supported by a structural wall in the center, while a handrail was inserted into the structural wall to save the width of the staircase handrail. Although the construction process could be cumbersome, we communicated our intentions and persuaded the contractor.

모따기99
Mottagi99
ⓒKyungsub Shin

거리에서 보이는 독특한 상부 모서리, 무거운 엉덩이를 받치는 듯한 아크 벽체 등에 적용된 '모따기'는 우리가 택한 하나의 방식으로, 이러한 환경에서 비롯된 경직된 경계들을 모호하게 만드는 시도였다. 모따기는 4, 5층 내부까지 연속해 변주되어 적용된다. 곡선의 틈으로 벌어진 공간에 계단이 놓였고, 머리 위로 채광을 들여 시간에 따라 그림자를 드리우는 곡선 천장이 만들어졌다. 건축적 경험뿐만 아니라 건축 형상적 아이덴티티를 만드는 것이 더욱 가치를 만들 수 있는 것을 이야기하며, 우리의 질문이었던 타이폴로지에 대한 도전을 상기했다.

모따기99
Mottagi99
ⓒKyungsub Shin

One of the methods we chose to apply to the unique upper edge when seen from the street, the arc which looks as if supporting heavy buttocks, was "chamfer," meant by the name "Mottagi." It was an attempt to blur the rigid boundaries resulting from this environment. The chamfer is applied in successive variations to the interior of the 4th and 5th floors. We placed a staircase in the interstice formed by the curve, and created a curved ceiling through which the sunlight comes, casting shadows differently depending on the time of day. Mentioning that not only the architectural experience but also the creation of architectural formative identity can further increase value, we recalled our question of challenging typology.

이처럼 밀집한 도심 필지에서의 공동주택 계획은 밀당의 연속이다. 세대수는 주차대수와, 필요 높이 확보는 일조제한선과 맞물려 어떤 것을 택하느냐에 따라 반대는 충족을 포기해야한다. 치열한 대립과 협의 속에 '무언가를 계획'한다는 것은 합당하고도 매력적인 이야기를 만들어내는 것이다. 아이디어와 콘셉트가 프로젝트를 시작하게 하는 단초라면, 내러티브는 프로젝트에 참여하는 클라이언트, 협력사, 시공사, 감리자, 허가를 담당하는 공무원에 이르는 모든 이에게 건축 계획과 의도, 그에 수반되는 것들을 설명하는 무기가 된다.

묵직한 엉덩이를 이상하게 떠받들고 있는 듯한 '모 딴' 덩어리는 우리의 흔한 주택가 동네에서 특별하고도 평범하게 앉아있다. 외장재로 붉은 벽돌을 적용한 것은 오랫동안 기존 동네를 구성해온 붉은 벽돌타일의 흐름을 이어간 것이다. 튀는 듯 튀지 않는 <모따기99>는 나름대로 자연스러운 방식으로 날카로운 질문을 던지고 있다.

Planning collective housing in such dense urban lots is a continuum of push and pull. The number of households is related to the number of parking spaces, and securing the necessary height is intertwined with the diagonal setback regulation for daylighting. So, whatever you choose, you may have to take one while giving up the other. "Planning something" through an intense consultation process is about creating a reasonable and attractive story. If the idea and concept are the starting point for initiating a project, the narrative is a weapon that explains the architectural plan, intent, and what it entails to everyone involved in the project, from clients, partners, contractors, supervisors, and government officials in charge of permits.

The "chamfered" mass that strangely seems to be supporting bulky buttocks occupies a place in an ordinary residential neighborhood in a special yet unique manner. The choice of red brick as exterior material continues the flow of red brick tiles that have long comprised the existing neighborhood. *Mottagi99*, which may stand out but also seems not to stand out, poses sharp questions in its own natural way.

모따기99 Mottagi99 ⓒKyungsub Shin

Agit Studio

텍토닉, 확장

2021년 5월, 국토부 인재육성사업에 선정되어 스위스에서 아지트 스튜디오 작업을 이어갈 기회를 만들 수 있었다. 개소 후 4년이 가까워지고 있었고 성수에서 혜화로 사무실을 옮긴 무렵, 사무실의 방향성과 지속성에 대해서 고민하던 단계였다. 공사단계 중인 프로젝트들은 하나 둘 완료되고 있었다. <프로젝트 양평>을 포함한 몇몇 작업이 계획단계에 있었지만, 코로나 여파로 비대면 회의가 자연스럽게 여겨지기 시작했던 그때는 한국을 잠시 떠나는 것이 클라이언트와의 소통을 멀게 만드는 것이 아니었기에, 충분한 동의와 응원 속에 스위스로 떠날 수 있었다.

리바 산 비탈레, 스위스
Riva San Vitale, Swiss
ⓒAGIT STUDIO

우리가 살았던 곳은 이탈리아 북부와 인접한 티치노 지역으로 리바 산 비탈레라는 작은 동네였다. 집 테라스에서 루가노 호수를 바라보고, 마리오 보타(Mario Botta)가 설계한 데너(마트)에서 장을 보고, 아울레리오 갈페티(Aurelio Galfetti)의 유치원과 체육관을 산책하며, 두리쉬 놀리(Durisch+Noli) 아름다운 공동주택 담장 앞에서 맥주를 마실 수 있는 곳이었다. 학생 때와는 다른 경제적 여건뿐 아니라 여행이 아닌 생활을 하게 된 이곳에서 흥미롭고 또렷한 시선으로 건축과 그 사회를 관찰할 수 있는 기회였다. 같은 사업의 해외연수분야에 선정된 허근일은 평소 관심을 가지고 있던 스토커 리 사무소(Stocker Lee Architetti)에서 근무했는데, 우리는 건축과 삶의 영역 모두에서 가족처럼 자리를 내어준 스토커 리와 그 주변 사람들 덕분에 더욱 깊고 적극적인 '로컬 이방인' 생활을 할 수 있었다.

Tectonic, Expansion

In May 2021, I was selected for the Ministry of Land, Infrastructure and Transport's talent development project and was given the opportunity to continue the work I was doing at Agit Studio in Switzerland. It had been nearly four years since Agit Studio opened, and it was around the time the office moved from Seongsu to Hyehwa. And I was pondering the direction and continuity of Agit Studio. Projects that were in the construction stage were all finished and there were several projects, including *Project YP*, that were in the planning stage. Since it was common to hold non-face-to-face meetings due to COVID-19, even if I left Korea for a specific period, I could still communicate with clients. Hence, people consented and supported me when I told them I would be leaving for Switzerland.

We lived in a small town called Riva San Vitale in Ticino canton adjacent to northern Italy. We gazed at Lake Lugano from the front of the house, shopped at Denner (supermarket) designed by Mario Botta, took walks through the kindergarten and sports center designed by Aurelio Galfetti, and drank beer in front of the fence of the beautiful collective housing designed by Durisch+Nolli. Not only was the economic situation different from when I was a student, but it was also an opportunity to observe architecture and society from an interesting and clear perspective in the place where I was living rather than traveling. Guenil Huh, who was selected for the overseas training category of the same project, worked at Stocker Lee Architetti, a firm he had always been interested in. Both Stocker and Lee and their friends welcomed us with open arms both professionally and personally and treated us like family. Thanks to them, we were "local strangers" deeply and actively indulged in the life there.

아지트 스튜디오 | Agit Studio

스위스에서 관심있게 보게 되는 것은 한 건축가가 자신의 이야기를 축적시키는 태도와 방식이었다. 작은 국가이지만 성격이 매우 분명한 각 지역을 기반으로 활동하는 스위스 건축가들이 평생에 걸쳐 반복적으로 확장하고 발전시킨 자신의 건축어휘들은, 시간을 버티는 힘이 있어 보였다. 그들은 건축을 '설계 혹은 디자인'한다기보다 건축가 각자의 '틀 혹은 영역'을 만들어 가는 듯 했다. 건축가의 구축적 내러티브가 구체적이고 집중적일 뿐 아니라 계속 확장되는 느낌을 받았기 때문이다. 특히 한 동네에서 건축가의 생애에 걸친 프로젝트들을 다양하게 볼 수 있었는데, 그것은 각 작업들 간에 존재하는 느슨하고도 질긴 연관성을 짐작해 볼 수 있는 바탕이 되었던 것 같다. 이 1여 년의 기간은 건축가로 일과 일상을 지속한다는 측면에서 익숙하고 자연스럽기도 했고, 빠르고 빡빡한 서울을 떠나 물리적, 사회적 밀도와 속도가 완전히 다른 곳에 있다는 점에서 새롭기도 한 생활이었다. 당시 가지고 있는 질문이나 경험에서부터 또 변화했을 것이기에, 이는 지금 아지트 스튜디오가 고민하는 부분들 – 동일한 구축 방식을 반복해서 써보고 발전시키고자 하는 것 등 - 에 대해 영향을 주었을 것이다. 언제가 이 시기 전후로 진행되었던 <프로젝트 양평>, <아카이브 매스>, <나주 주택> 등을 이 같은 시각으로 관찰해보는 것도 의미 있을 것 같다.

프로젝트 양평
Project YP
ⓒPHSG

What interested us in Switzerland was the attitude and method of an architect accumulating his story. There were architects based in each region with very distinct characteristics, and the architects' respective architectural vocabularies, which were repeatedly expanded and developed throughout their lives, seemed to have the power to withstand time. Each architect seemed to be creating one's own "framework or area" rather than "designing" architecture. This is because I felt that the architect's tectonic narrative was not only specific and focused, but also continuously expanding. In particular, it is possible to see a variety of projects spanning the architect's life in one neighborhood, which became the basis for my conjecture of the loose yet tenacious connections that exist between each work. The period of about a year was familiar and natural in the sense that I continued my work and daily life as an architect, but also allowed me to enjoy a new life in a place with completely different physical and social density and pace, leaving behind life in the fast and busy city of Seoul. The questions and experience we had at the time would probably be different if we underwent the same process today. Even so, those questions and experiences would have had an impact on the areas that Agit Studio is currently concerned about, such as repeatedly using the same construction method and trying to further develop it. Going forward, it would be meaningful to observe projects such as *Project YP*, *Archived Mass* and *Naju Housing Project* that were conducted around this time from this perspective.

Agit Studio

<프로젝트 양평>의 볼륨을 달리하여 나눈 매싱(massing)은 층으로 구분된 것이 아니다. 하나의 덩어리로 보이고 싶었던 외부는 같은 종류지만 온장, 반장, 삼분의 일장, 세 가지 크기로 가공된 벽돌이 적용됐다. 수많은 벽돌 중에서 선택된 것은 크기나 모양을 다양하게 가공한 콘크리트 벽돌이나 말끔하게 조색된 벽돌이 아닌, 그야말로 흙으로 구워 만든 벽돌이었다. 의도하는 흙색을 찾기 위해 중국에서 고벽돌을 가져왔다. 개개의 벽돌들이 정직하게 집합되는 가장 기본적인 쌓기방식을 취하는 것도 오직 전체로 느껴지는 효과를 강조하기 위해서였다. 사람들이 걸어 다니며 볼 수 있는 제일 낮은 부분에 가장 작은 크기의 벽돌을 쌓았고, 지상에서 멀어지는 상층부는 규칙에 따라 더 크게 가공된 벽돌을 적용하여 원근감과 시공 편의성을 고려했다. 분할된 각 매스는 하부 매스보다 5cm씩 더 바깥으로 돌출되었다. 이 돌출된 하부는 별도의 철물장치를 받치지 않고 벽돌로서 오롯이 보일 수 있도록 디테일을 고민했고, 그로 인해 생기는 그림자는 단일한 질감으로 건물 볼륨을 풍성하게 한다. 모두 같은 색으로 표현된 건물의 금속 부분들 또한 벽돌로 이루어진 '덩어리'를 방해하지 않는 역할을 하고 있다. 창호 프레임, 핸드레일, 캐노피, 물 홈통은 각각 알루미늄, 스틸, 스테인리스 등 다른 종류의 금속이지만, 최대한의 노력을 들여 가공하여 단일한 색으로 만들었다. 우리 작업에서는 '디테일'이라고 불리는 모든 것이 '덩어리'라는 주제를 위해 수렴된다.

프로젝트 양평
Project YP
ⓒPHSG

The massing divided into different volumes in *Project YP* is not distinguished by floors. We wanted the exterior to look like a single mass, and thus used bricks that are produced using the same processing method but in three sizes: full brick, half bat, and one-third bat. Among the numerous types of bricks, the one we chose was not concrete brick with assorted sizes or shapes nor neatly colored bricks, but bricks made by baking soil. To find the right earth color, we brought in old bricks from China. The most basic brick-laying method, in which individual units are assembled in an honest manner, was applied for the sole purpose of emphasizing the whole effect of bricks. The smallest bricks were laid at the lowest part of the exterior that people walking by can touch, and larger bricks were stacked according to a set rule at the upper levels, going farther away from the ground, to cater to the perspective and ease of construction. Each of the divided mass protrudes outward by 5 cm more than the mass beneath. We gave careful thought to the details so that there is no separate steel apparatus supporting the lower part of the protruding section and that only bricks are visible. As a result, when a shadow is cast, the building's volume is enriched with a single texture.
All metal parts were painted the same color as the bricks to prevent them from interfering with the "mass" made of bricks. Even though the window frames, handrails, canopy, and gutters are made of different metals such as aluminum, steel, stainless steel, etc., utmost efforts were made to process them so that a single color was expressed. In our work, all things referred to as "details" center around the theme "mass."

프로젝트 양평
Project YP
ⓒKyungsub Shin

Attitude of AGIT STUDIO | 태도; Attitude

프로젝트 양평
Project YP
ⓒKyungsub Shin

아지트 스튜디오 | Agit Studio

코어의 존재와 위치가 이형으로 길게 놓인 대지에서 전체 건물의 배치, 동선, 레벨을 결정한다. <프로젝트 재해석>과 마찬가지로 <프로젝트 양평>에서도, 사용자를 미리 알 수 없는 근생 건물에서 '코어'가 담당하는 역할에 대한 우리의 생각이 드러난다. 계단실, 로비, 엘리베이터, 화장실 등 서번트부가 모인 코어는 정체성을 만드는 주요부다. 상부에서 들어오는 빛이 잘 머물 수 있도록 질감과 구성을 계획했다. 외부에서 경험한 감각들이 내부의 경험으로도 이어질 수 있기를 바랐다.

프로젝트 양평
Project YP
ⓒPHSG

처음, <프로젝트 양평>은 부지의 조건을 뒤집을 수 있을 것인가라는 질문에서 시작했다. 6차선과 2차선 도로가 만나는 코너로 차와 사람들이 많이 오가는 랜드마크적 성격을 가진 곳임에도 불구하고, 건폐율 20%의 대지 조건과 층수 제한(4층)이 만드는 제약을 반전시키고 싶은 의도였다. 한 층 바닥면적이 90㎡에 불과한데 어떻게 코너 건물의 존재감을 드러내고 사람들에게 공간적 경험을 제공할 것인가, 이형의 대지를 지배력 있게 사용하고 규모를 확보하는 지하층을 계획한다면 빛과 동선을 어떤 식으로 정리해야 할 것인가, 층이라는 형상학적인 구분을 지우고 '매스' 자체로서 랜드마크가 되게 하는 방법은 무엇일까⋯⋯ 우리는 건축물을 이루는 다양한 개념을 일관된 하나의 구축어휘를 가지고, 그것을 변주시켜 해결하고자 하는 생각을 지속했다.

Attitude of AGIT STUDIO

The presence and location of the core determined the layout, circulation, and level composition of the entire building on a long piece of land with uneven shape. Like *Project Re-interpret*, *Project YP* is also a retail building whose users cannot be known in advance, which reveals the role of the "core" of the building. The core, which consists of servant parts such as the staircase, lobby, elevator, and restrooms, is the main part that creates identity. Its material texture and composition were planned so that the light coming in from the top can continue. That is, we hoped that the external experience of senses can lead to the internal experience.

Project YP began with the question of whether the conditions of the site could be overturned. The site is located on a corner where 6-lane and 2-lane roads meet and thus possesses characteristics of a landmark where many cars and people pass by, but regardless of such conditions, it is restricted to the building-to-land ratio of 20% and to the height of 4 floors from the ground up. The intent was to overturn such restrictions. When the area of one floor is only 90 sq. m., how can we reveal the presence of a corner building and how can we provide spatial experience to people? If the plan is such that the land with uneven shape is used with dominance, and some scale is secured on the basement level, then how can we manage light and circulation? How can we erase the morphological division of floors and make the "mass" itself a landmark? We continued contemplating these questions through massing by transforming the various concepts that make up a building into a single construction vocabulary.

프로젝트 양평
Project YP
ⓒKyungsub Shin

음악을 그리는 악보 오선 위의 점 하나가 허투루 찍히지 않듯, 수백장 도면으로 옮겨지는 우리 세계를 옆구리에 끼고 다니며 바닥도 없었던 대지 위에 하나의 영구한 덩어리와 새로운 관계들을 만든다. 많은 회의와 복잡한 조율 과정, 눈앞에 보이는 공사 과정의 수고보다도 보통 더 길고 진하기 마련인 계획의 시간 동안, 오롯이 치고받고 풀어내고 그려내고 쓰고 지우고 만드는 과정이 우리의 악보이자 글이자 시나리오인 도면 위에서 수렴되고 정의된다. 그 생각의 힘이 정확하고 분명할수록 실현은 가까워진다. 건축을 만드는 시간은 이렇듯 생각을 명료히 다듬어가는 과정 자체라고 생각한다. 아지트 스튜디오의 이야기는 질기고도 느슨하게 방향성을 가지고 엮어가고 있다. 그리고 그 위에 자발적 혹은 우발적으로 생기는 여러 외부 자극들이 우리의 가능성과 능력을 훨씬 더 확장하고 증폭시켜준다. 그것이 우리가 지금까지의 이야기보다 앞으로의 이야기를 기대하며, 더 크게 나아갈 수 있는 이유이자 동력이다.

프로젝트 양평
Project YP
ⒸKyungsub Shin

Just as due consideration is given in placing even a single dot on the musical staff, we transfer our world to hundreds of drawings, carry them around and create a new relationship between the previously empty land and a permanent mass. Time spent on planning is by and large longer and more intense than the many meetings, complicated coordination process, and the visible efforts of construction process. Processes of heated exchange, resolving issues, drawing, writing, erasing, and creating are converged on and defined in the drawings, which is our musical staff, our writing, and our scenario. The more accurate and clear the power of the thought, the closer it comes to realization. In this way, I believe that the time of creating architecture is the process of refining one's thoughts clearly. The story of Agit Studio is tightly and loosely woven in a certain direction. And on top of that, various external stimuli that occur voluntarily or incidentally expand and amplify our possibilities and abilities even further. That is the reason and driving force that allow us to move forward even further, with high expectations for the future story to surpass what has transpired so far.

프로젝트

콘크리트 도서관
- 위치: 부산광역시 금정구 장전동 222-57
- 용도: 도서실(근린생활시설), 단독주택
- 책임건축가: 서자민, 허근일
- 대지면적: 220.5m²
- 건축면적: 113.50 m²
- 연면적: 198.51 m²
- 건폐율: 51.47%
- 용적률: 90.03%
- 규모: 지상 2층
- 구조: 연와조, 철근콘크리트
- 설계기간: 2018. 01. ~ 2020. 01.
- 시공기간: 2019. 01. ~ 2020. 01.
- 구조설계: ㈜인 구조안전기술
- 시공: ㈜채헌종합건설
- 사진: 신경섭

프로젝트 재해석
- 위치: 부산광역시 금정구 장전동 418-44, 45
- 용도: 근린생활시설
- 책임건축가: 서자민, 허근일
- 대지면적: 419.10m²
- 건축면적: 242.27m²
- 연면적: 1,446.36m²
- 건폐율: 57.80%
- 용적률: 282.50%
- 규모: 지하 1층, 지상 5층
- 구조: 철근콘크리트, 철골
- 설계기간: 2019. 03. ~ 2021. 03.
- 시공기간: 2020. 09. ~ 2021. 07.
- 구조설계: ㈜이든구조컨설턴트
- 기계, 전기설계: 한누리엔지니어링
- 시공: ㈜채헌종합건설, ㈜엔원종합건설
- 사진: 신경섭

아카이브 매스
- 위치: 서울특별시 서대문구 연희동 719-10
- 용도: 갤러리(근린생활시설), 단독주택
- 책임건축가: 서자민, 허근일
- 대지면적: 93.59m²
- 건축면적: 46.40m²
- 연면적: 180.32m²
- 건폐율: 180.32m²
- 용적률: 146.13%
- 규모: 지하 1층, 지상 3층
- 구조: 철근콘크리트, 연와조
- 설계기간: 2022. 02. ~ 2022. 07.
- 시공기간: 2022. 07. ~ 2022. 11.
- 기계, 전기, 통신설계: 아이에코 ENG
- 시공: ㈜제이종합건설
- 사진: 조엘 모리츠

모따기99
- 위치: 서울특별시 강동구 명일동 99
- 용도: 다세대주택, 근린생활시설
- 책임건축가: 서자민, 허근일
- 대지면적: 218.30m²
- 건축면적: 128.02m²
- 연면적: 436.01m²
- 건폐율: 58.64%
- 용적률: 199.73%
- 규모: 지상 5층
- 구조: 철근콘크리트, 철골
- 설계기간: 2019. 04. ~ 2019. 11.
- 시공기간: 2019. 11. ~ 2020. 07.
- 구조설계: ㈜이든구조컨설턴트
- 기계, 전기, 통신설계: 아이에코 ENG
- 시공: 다미건설(주)
- 사진: 신경섭

프로젝트 양평
- 위치: 경기도 양평군 양평읍 공흥리 722-2
- 용도: 근린생활시설
- 책임건축가: 서자민, 허근일
- 대지면적: 455.0m²
- 건축면적: 90.9m²
- 연면적: 499.87m²
- 건폐율: 19.98%
- 용적률: 77.20%
- 규모: 지하 1층, 지상 4층
- 구조: 철근콘크리트
- 설계기간: 2021. 03. ~ 2022. 06.
- 시공기간: 2022. 06. ~ 2023. 06.
- 구조설계: ㈜이든구조컨설턴트
- 기계, 전기, 통신설계: 아이에코 ENG
- 시공: ㈜제이종합건설
- 사진: 신경섭, PHSG

Projects

CONCRETE LIBRARY
- Location: Geumjung-gu Janjun-dong 222-57, Busan, Republic of Korea
- Program: Library & Single-family House
- Architects: Jamin Seo, Guenil Huh
- Site area: 220.5m²
- Building area: 113.50 m²
- Gross floor area: 198.51 m²
- Building coverage ratio: 51.47%
- Floor area ratio: 90.03%
- Number of levels: 2F
- Structure: Brick structure, Reinforced concrete
- Design Period: 2018. 01. ~ 2020. 01.
- Construction Period: 2019. 01. ~ 2020. 01.
- Structural engineering: IN Structure
- Construction: Cheahun construction
- Photography: Kyungsub Shin

Project Re-interpret
- Location: Geumjung-gu Janjun-dong 418-44, 45, Busan, Republic of Korea
- Program: Commercial & Office
- Architects: Jamin Seo, Guenil Huh
- Site area: 419.10m²
- Building area: 242.27m²
- Gross floor area: 1,446.36m²
- Building coverage ratio: 57.80%
- Floor area ratio: 282.50%
- Number of levels: B1, 5F
- Structure: Reinforced concrete, Steel structure
- Design Period: 2019. 03. ~ 2021. 03.
- Construction Period: 2020. 09. ~ 2021. 07.
- Structural engineering: Eden Structural Consultant
- Mechanical, Electrical Engineering: Hannuri Engineering
- Construction: Cheahun Construction, N1 Architecture
- Photography: Kyungsub Shin

ARCHIVED MASS
- Location: 97, Hongyeon-gil, Seodaemun-gu, Seoul, Republic of Korea
- Program: Gallery & Single-family house
- Architects: Jamin Seo, Guenil Huh
- Site area: 93.59m²
- Building area: 46.40m²
- Gross floor area: 180.32m²
- Building coverage ratio: 180.32m²
- Floor area ratio: 146.13%
- Number of levels: B1, 3F
- Structure: Reinforced concrete, Brick Structure
- Design Period: 2022. 02. ~ 2022. 07.
- Construction Period: 2022. 07. ~ 2022. 11.
- Mechanical, Electrical, Communication Engineering: IECO ENG
- Construction: J Construction
- Photography: Joel Moritz

MOTTAGI99
- Location: Gangdong-gu Myeongil-dong 99, Seoul, Republic of Korea
- Program: Multi-family housing & Retail
- Architects: Jamin Seo, Guenil Huh
- Site area: 218.30m²
- Building area: 128.02m²
- Gross floor area: 436.01m²
- Building coverage ratio: 58.64%
- Floor area ratio: 199.73%
- Number of levels: 5F
- Structure: Reinforced concrete, Steel structure
- Design Period: 2019. 04. ~ 2019. 11.
- Construction Period: 2019. 11. ~ 2020. 07.
- Structural engineering: Eden Structural Consultant
- Mechanical, Electrical, Communication Engineering: IECO ENG
- Construction: Dami Construction
- Photography: Kyungsub Shin

PROJECT YP
- Location: 64-4, Jungang-ro, Yangpyeong-eup, Yangpyeong-gun, Gyeonggi-do, Republic of Korea
- Program: Commercial & Office
- Architects: Jamin Seo, Guenil Huh
- Site area: 455.0m²
- Building area: 90.9m²
- Gross floor area: 499.87m²
- Building coverage ratio: 19.98%
- Floor area ratio: 77.20%
- Number of levels: B1, 4F
- Structure: Reinforced concrete
- Design Period: 2021. 03. ~ 2022. 06.
- Construction Period: 2022. 06. ~ 2023. 06.
- Structural engineering: Eden Structural Consultant
- Mechanical, Electrical, Communication Engineering: IECO ENG
- Construction: J Construction
- Photography: Kyungsub Shin, PHSG

리뷰   태도를 잃지 않는 날것의 감동   양수인

아지트 스튜디오
2023년 젊은 건축가상 수상자인 아지트 스튜디오는 대표 서자민이 운영하고 있다. 젊은 건축가상 심사를 하며 처음 알게 되었고, 그 이후 이 글을 쓰기 위해 몇 번이나 만나 이야기하면서 조금 더 알게 되었다. 아지트 스튜디오의 작업에 대한 논의는 서자민 건축가와 파트너로서 영향을 주고받으며 함께 활동하고 있는 허근일 두 건축가에 대한 이야기임을 밝히며, 아지트 스튜디오 작업에 대해 몇 가지 느낀 점을 적어본다.

명확한 '태도'
우리 엄마가 이해 못 하실 이야기는 하지 말자.
    온갖 현학적인 이야기를 늘어놓거나, 불명확한 그림을 펼치고 소위 말해서 썰을 풀기 시작하는 직원들이나 학생들한테 종종 하는 말이다. 다양한 요소들이 뭔가 있어 보이기는 하는데 하나로 잘 뭉쳐지지 않아 핵심을 파악하기 힘들거나, 지나치게 거시적인(과대망상적인) 관점의 작업을 대할 때도 외칠 수밖에 없는 문장이다.
    물론 건축적 논의가 하향 평준화되거나 대중성을 지나치게 지향해야 하는 것은 아니지만, 건축설계는 근본적으로 건축물에 대해 생각하고 실현하는 실용적인 서비스업이니 (실험적, 철학적, 이론적인 건축일지라도) 우리 엄마도 쉽게 이해할 만큼 명확하게 구현하고, 명료하게 표현할 수 있을지 여부는 나에게 중요한 하나의 지표이다.
    아지트 스튜디오의 작업은 우리 엄마도 참 잘 이해할만한 건축이라고 생각했다. 서자민이 글에서 굳이 작은따옴표를 붙여 표기하는 건축적 '태도'는 그들의 모든 작업에서 단순하리만큼 명확하다. 각 작업들은 그것이 처한 특정한 상황에 대해 건축가가 오랜 시간 고민하면서 서서히 무르익은 것이겠지만, 결국 어느 순간 간단한 명제로 표현된다. 때로는 굉장히 직관적으로 느껴지기도 한다. <콘크리트 도서관>이 특히 그러하다. 신축이 불가한 땅에 지반침하가 일어나고 있는 조적조 건물을 살리기 위해 조적벽과 콘크리트벽을 더한 두꺼운 하이브리드 구조벽을 만든다는 아이디어는 건물의 경험과는 별개로 듣는 순간 이미 고개를 끄덕이게 한다. 피할 수 없는 조건은 최대한 받아들이고 오히려 부각해 장점으로 만든다는 단순하고도 명확한 건축적 태도가 그 작업이 처한 상황과 딱 맞아떨어졌을 때 (관객으로서 또는 동종업계 종사자로서) 일종의 쾌감을 느끼게 된다.
    건축뿐 아니라 모든 창작과정은 사실 근본적으로 그 태도를 명확히 하는 과정이다. 작업을 대하는 태도가 명확하면, 설계 및 시공과정에서 접하게 되는 수없이 많은 선택의 순간들은 의외로 스스로 답을 보여준다.

Review
Moved by the Raw Unwavering in Attitude

Sooin Yang

Agit Studio

Agit Studio, one of the Korean Young Architect Award 2023 winners, is run by Principal Jamin Seo. Since I first knew this studio by joining the KYAA jury, I met them several times to write this essay and came to know a little more of the studio. Making it clear that this essay on Agit Studio's works focuses on Jamin Seo and the visiting partner Guenil Huh who is collaborating and exchanging influences with her, I will narrate the several points I felt from the studio's works.

A Clear "Attitude"

"Let's not talk about what our mothers could not understand."

This is what I say to some employees or students who start to talk all kinds of pedantic stories or unfold a narrative about their indefinite drawings. I cannot but say so when some various elements seem to be found but not so well combined that it is hard to figure out their core idea, or when a presentation of work I see renders its perspective too macroscopic or grandiose.

Although it would not be desirable for architectural discussions to be levelled downward or oriented to popular appeals too much, architectural design is basically a practical service to consider and realize buildings so that it is one of my important touchstones whether to render as clearly and express as lucidly as even my mother could understand.

I found Agit Studio's works to be such an architecture understood very well even by my mother. The architectural "attitude," emphasized by Seo with quotation marks, is as clearly revealed as even feeling simple in all of their works. Each work would have gradually been ripen through the architects' long considerations on the specific contexts, but at some moments they are expressed as simple propositions. Sometimes they feel very intuitive. A case in point is *Concrete Library*. Apart from its experience, I just made a nod upon hearing the idea of creating the thick hybrid structural walls combining masonry with concrete walls to save the masonry building which is undergoing ground subsidence on a land where new constructions are impossible. When their simple but clear architectural attitude to accept inevitable conditions as many as possible and emphasize them as advantages matches well with the situation of their works, I feel a kind of pleasure (as an observer or a peer practitioner in the same profession).

Not only architecture, but every creative process is basically the process of clarifying such an attitude. If the working attitude is clear, a countless number of choices faced during the design and construction phases reveal answers by themselves.

진술한 의도

나는 프로젝트 런웨이나 넥스트 인 패션 같은 다른 크리에이티브 업계의 경연 TV 프로그램을 좋아한다. 약간 비슷하지만 너무나도 다른 사고 과정을 보는 것이 신기하고 재미있기 때문이다. 패션업계의 디자인 과정을 보면 '영감'을 얻어 시작을 하는 경우가 많다. 그런데 그 '영감'이 어디에서 왔는지 도저히 이해할 수 없는 경우가 대부분이다. 밑도 끝도 없이 왜 1930년대의 재즈에서 영감을 받는지 모르겠다. 물론 대다수의 경우 그 영감에 의해 옷은 아름답게 탄생하므로 사실 영감이 어디에서 기인하건 큰 상관이 없을 것이다. 건축계에서는 오히려 반대의 경우를 많이 보게 된다. 커리어 전반을 아우르는 일관성 있는 어젠다를 실험하며 자신의 작업을 하나의 개념으로 묶으려고 하는 경우, 어떻게 모든 의뢰인의 모든 땅에 같은 개념이 적용 가능할까 의문이 드는 경우가 있다. 약간 억지스럽게 느껴진달까? 너무 애를 쓰는 것 같아 보인달까? 각 작업에 대한 아지트스튜디오의 출발점은 굉장히 진솔하다.

굳이 진솔하다는 표현을 쓴 이유는 각 작업의 출발점이 지극히 현실적이고 실용적으로 진실하기 때문이다. 또 겉으로 그럴싸해 보이는 포장을 하지 않고 솔직담백하게 풀어나가기 때문이다. 쉽게 이해하고 수긍할 수 있다.

<프로젝트 재해석>은 외부에서 디자인 의도를 끌어오는 것보다 건축가에게 요구된 실용적 건축행위, 즉 새롭게 풀어내야 했던 계단실과 서번트 공간들을 재해석해 만들어내는 작업을 적극적으로 드러내고 해당 건물 및 가로에 조금은 새로운 생명력을 불어넣은 작업이다.

<프로젝트 양평>의 경우 건폐율 20%의 한계를 극복하기 위해 대지 고저차를 활용해 지하공간을 극대화하고, 건물 코어가 극적인 공간 경험이 되며 건물 자체가 하나의 큰 덩어리로서 매력을 갖도록 진솔한 건축해법들을 조합했을 뿐이다.

치밀한 실행

태도가 전쟁이고 의도가 전략이라면, 실행은 전술이다. 젊은 건축가는 장군이기도 하고 동시에 보병이기도 한 것이다.

2년 반 동안 현장을 왔다 갔다 진두지휘하면서 완성한 <콘크리트 도서관>의 실행 과정은 물론 치열했겠으나, 어떤 연령대의 건축가에게도 현장의 치열함은 당연한 과정이므로 논외로 하겠다. 사실 건물은 현장에서 물리적으로 실행되기 이전에 건축가의 생각 속에서 종이 위의 선으로 먼저 실행된다. 아지트 스튜디오의 치밀한 실행이 독특한 빛을 발하기 시작하는 순간은 현장에 가기 이전, 생각이 종이 위에 모습을 드러내는 시점이다. 그들 스스로가 여러 글에서 언급한 '못생김'이 바로 치밀함과 연결되는 지점이기도 하다. 사실 사유의 영역인 태도나 의도를 물질화하는 건축설계 과정에는 수많은 유혹이 존재한다. 형태는 기능을 따른다는 생각은 무너진 지 오래이고, 첫 완공작을 선보이기 이전부터 인스타그램 등의 매체를 통해 자신을 기가 막히게 어필하는 '더 젊은' 건축가들이 많은 시대에 '못생김'을 이야기하기는 쉽지 않다. 물론 일부러 못나게 만들자

Honest Intentions

I enjoy such TV survival shows in other creative industries as Project Runway or Next in Fashion, because it is amazing and fun to see the similar but too different thinking processes. In fashion industry, the design process often begins with an "inspiration," the origin of which, however, is hardly understandable in most cases. There is no way of understanding why such a designer was inspired by the jazz of the 1930s out of nowhere. Given such a clothing is born as beautiful thanks to such an inspiration in most cases, where it came from might not be such a big deal. This is contrary to the most architectural scenes. When a consistent agenda covering an architect's overall career was experimented on to bind his or her own works into a concept, I would wonder how one and the same concept could apply to all the clients' lands? Does it not a little farfetched? Is it not endeavoring too much only to look contrived? In contrast, Agit Studio starts their works from a very honest motive.

The reason why I used the expression "honest" is that the starting point of each work is extremely realistic and practically genuine. Also, it is not speciously covered but frankly and forthrightly unfolded, so as to be easily understood and agreed on.

Project Re-interpret is a work that actively revealed the architect's reinterpretation of the client's practical requirements as designing the staircase and servant spaces in a new way, rather than borrowing a design intention from outside, and so inspired a little new vitality to the building and the street.

For *Project YP*, the architect just used the level difference of the site to maximize the underground space so as to overcome the limitation of 20% building coverage, combining honest architectural solutions with the building core providing a dramatic spatial experience and the building itself having a charm as one large mass.

Meticulous Implementations

If the attitude is the war, and the intention is the strategy, the implementation is the tactics. Young architects are both generals and foot soldiers.

The implementation process of *Concrete Library* completed for two and a half years would have surely been intense, but this intensity on site is a matter of course for any age of architects and so left out of my discussion here. Actually, a building is implemented in the architect's mind as the lines on a paper before it is physically realized on site. Agit Studio's intense implementation starts to give off a unique glow at the moment when their ideas are revealed on paper before they visit the project site. At this point in time, the "ugliness" they mentioned in many writings is also connected to such intensity. Indeed, a number of lures are in the architectural design process of materializing the attitude or intention, the realm of thinking. Given it is a long time since the idea "Form follows function" collapsed, and in this era when many "younger" architects who ravishingly appeal themselves through media like Instagram even before showing their first completed work, it is not easy to discuss the "ugliness." This would surely not mean any intentional argument for creating ugliness or any indifference to beauty. Rather, it would be a self-censorship device to prevent the lure of

거나 아름다움에 관심이 없다는 것은 아닐 것이다. 다만 그것은 예쁨의 유혹이 태도와 의도를 흐트러트리지 못하게 하는 하나의 자기검열장치일 것이다. 소위 말해 '인스타그래머블'한 색감이나 화려하고 유혹적인 형태보다는 '덩어리'와 '텍토닉'에 먼저 집중해 보겠다는 선언이기도 하리라.

<아카이브 매스>는 치밀한 절제미를 잘 보여준다. 특별하거나 눈에 띄는 재료를 배제한 무덤덤한 모습은 소위 페인트칠 좀 다시 했나 보네라고 생각할 만한 외관이다. 여기에 특별함을 더하는 요소는 보일 듯 말 듯 남겨진 창호의 원형이다. 하얗고 평평한 갤러리의 내부에 특별함을 더하는 요소는 원래 있던 삐뚤빼뚤함과 울퉁불퉁함이다.

젊은 건축가상
젊은 건축가상 심사를 맡으며 수상 후보로 세 부류를 생각해 보았다.

먼저 '젊은' 계급장을 떼고도 매우 완성도가 있는 작업을 하는 건축가이다. 응당 수상해야 한다고 생각했다.

두 번째로 지속가능하고 확장가능한 시스템을 만들어가고 있는 건축가이다. 당장 앞에 주어진 프로젝트를 잘 설계하고 실현하는 것 이상으로 자신의 건축, 또 사무실의 운영에 있어서 앞으로도 긍정적인 방향으로 발전할 수 있는 시스템을 스스로 만들어가고 있는 젊은 건축가도 수상감이라고 생각했다.

마지막으로 가장 기대 되고 젊은 건축가상의 취지에 맞는다고 생각한 경우는 규격화되기 이전 날것의 상태가 신선하게 남아있는 건축가이다. 건축가로 성장하고 사무소도 규모가 커지게 되면 생존을 위해 어느 정도의 효율화가 수반되어야 하는데, 이는 곧 규격화를 의미한다. 사무실 운영의 규격화, 디테일의 규격화, 감리 밀도의 규격화 등, 젊은 건축가의 입장에서 이 규격화 이전의 정리되지 않은 상태는 필연적인 출발선이며 진정한 젊음의 기회이기도 하다.

크게 크게 휘두르는
건축가 서자민은 젊은 건축가상의 취지에 가장 들어맞는, 규격화되기 이전의 신선함이 남아 있는 건축가라고 생각했다. 응모한 대다수의 포트폴리오와 마찬가지로 아지트스튜디오의 포트폴리오도 소규모 건물이 주를 이루는데, 특별히 모자라지 않았지만 그렇다고 아주 눈에 띄지 않는 것도 사실이었다. 개별 작업의 완성도가 뛰어났다기보다는 각 작업에서 필요한 문제를 명확히 정의하고 스스로 정의한 문제를 해결해 나가는 방법이 과감하고 신선했다. 하지만 가장 눈에 띄고 기억에 남았던 것은 서자민이 자신의 작업에 대해 이야기하는 방식이었다. 포트폴리오를 보았을 때와 달리, 실제 대면해서 이야기를 듣고 질의응답을 하면서 가장 안정감을 느꼈다. 강단 있는 건축가라고 느꼈는데, 어떤 특정한 건물에 수여하는 상이 아니고 건축가에게 수여하는 상이라는 측면에서 수상할 만하다고 생각했다.

나는 건축가로서 진짜 순수하게 건축적인 유일한 순간은 그 건축을 맥락화시키는

prettiness from disturbing the attitude or intention. In other words, it would also be a declaration to focus early on "massing" or "tectonic" rather than the so-called "Instagrammable" colors or the fancy and luring forms.

Archived Mass is a good case of representing the beauty of elaborate moderation. The placid look excluding any special or eye-catching material is the facade worthy of looking as if repainted. An element adding uniqueness here is the original form of the windows left to be scarcely perceivable. The original crookedness and bumpiness add uniqueness to the inside of the white flat gallery.

The Young Architect Award
Participating as a jury member of the KYAA 2023, I thought about three kinds of prospect winners.

First, an architect who shows high-quality works regardless of the "young" insignia: this kind of architect is thought to be definitely worth a winner.

Second, an architect who is creating a sustainable and expandable system: this young architect who is creating a system developable in a positive direction in terms of one's own architecture as well as its practice operation, beyond just managing to design and realize given projects at hand, is also thought to be worth a winner.

Finally, an architect whose raw and unstandardized aspects remain fresh is thought to be most promising as fit for the intent of the KYAA. For one who grows into a more established architect operating a larger-scale practice, a certain degree of streamlining, which is to say, standardizations should be accompanied. The unorganized state before the standardizations – those of practice operation, details, supervision profundity and the like – is the necessary starting line for a young architect as well as the real chance of youth.

Daring and Bold Strokes
I thought Jamin Seo was the fittest architect for the intent of the KYAA, an architect who still has a sense of freshness before standardizations. Like the majority of applicants' portfolios, Agit Studio's portfolio primarily comprises small-scale buildings. While not particularly lacking in quality, they were obviously not quite eye-catching. Rather than excelling in the quality of individual works, the bold and innovative aspect lay in clearly defining the issues required for each project and solving the self-defined problems. However, the most eye-catching and memorable was how Seo narrated the story of her works. Unlike when I first looked at the portfolio, I felt the most reassured during the face-to-face interaction, listening to the presentation and engaging in a question-and-answer session. I felt she was a strong-minded architect worth to be a winner of this award, which is presented not to a certain building but to an architect.

As an architect, I believe the really purely architectural moment is only the moment of contextualizing the architecture. Architects, the professionals who have to design others' buildings with others' money, cannot decide and behave at our disposal: this is the very profession's limitation and attraction. Really at

순간이라고 생각한다. 기본적으로 남의 돈으로 남의 건물 지어주는 직업인지라 온전히 내 마음대로 판단하고 행동할 수만은 없는 것이 건축가라는 직업의 한계이자 묘미이다. 정말 내 마음대로 할 수 있는 것은 내가 설계한 건물이 이 사회에, 도시에, 건축사에 어떻게 자리매김할 것인지 그 맥락을 정하고 이야기하는 순간이다. 단순 건축업자를 넘어선 건축가이고 싶다면 자신의 건축에 대한 생각을 동료 건축가들과 나누고 평을 받기 위해 필수적인 작업의 과정이라고 생각한다. 이 작업은 누구에게 보수를 받고 하는 일이 아니기에 완전히 자유로울 수 있지만, 또한 온전히 내 책임이기도 하다.

적당한 규모의 건물 도면과 사진을 보여주면서 서자민이 들려준 이야기는 지극히 현실적이기도 했지만, 동시에 항상 조금 더 큰 관점을 이야기하고 있었다. <모따기99>는 모따기라는 형태적 어휘의 변주를 통한 밀도의 재배치, 건축적 조형성에 대한 실험이기도 했지만 스스로 다세대주택이라는 도시건축 유형에 대한 도전이라고 선언했으며, <콘크리트도서관>, <프로젝트 재해석>, <아카이브 매스> 역시 예산이 한정된 조건에서의 실용적인 리모델링 작업임과 동시에 동일한 지역지구 내에 비슷한 시기 지어진 비슷한 형태의 건물의 지속가능한 전략을 제안하는 작업으로 자리매김했다. 각 작업에 이미 충분히 담겨있는 특수해를 넘어 추상화를 통해 확장가능한 일반해를 제시하는 모습은, 에피소드의 나열을 통해 아름다움에 대해 이야기하고 내적인 질서를 통해 치밀한 완성도를 만들어가는 과정을 보여준 다른 두 수상자와 흥미로운 대비를 이루었다. 조심스레 정진하기보다는 일단 크게 크게 휘두른다는 생각이 들었는데, 젊은 건축가의 패기라고 생각됐다.

명확하고 진솔하고 치밀한 것은 이해할 수 있고 공감할 수 있다.
하지만 큰 감동은 종종 이해 불가한 것에서 오기도 한다.

양수인
양수인은 서울을 기반으로 활동하는 디자이너이다. 건축, 참여적 예술, 디자인, 마케팅, 브랜딩 등 광범위한 영역에서 건물, 공공예술, 체험 마케팅, 손바닥만 한 전자기기, 단편영화까지 다양한 스케일과 매체로 작업한다. 다양한 매체를 통한 디자인 작업이 모두 직면한 과제를 의뢰인의 상황에 부합하는 형식으로 해결하는 과정으로서 근본적으로 크게 다르지 않은 행위라고 믿으며, 그 근저에는 어떤 '것'을 만듦으로써 '삶'에 긍정적인 영향을 줄 수 있는 이야기를 전달하고자 하는 공통적인 목표의식을 갖고 작업한다.

양수인은 연세대학교 건축공학과와 뉴욕 컬럼비아 건축대학원 졸업 후, 이례적으로 졸업과 동시에 컬럼비아 건축대학원 겸임교수 및 리빙 아키텍처 연구소장으로 7년간 재임했다. 2011년 서울에 돌아와 삶것/Lifethings 라는 조직을 꾸려 활동하고 있다.

our disposal is just the moment when we set and discuss the context of how the buildings we designed will settle in this society, this city, and the history of architecture. I believe it is the essential process of a work to share one's own ideas and feedbacks with peer architects, if we are to transcend simple commercial builders. This work is not to be paid for by someone, so as to be completely free, but also my whole responsibility.

Presenting decently scaled building drawings and photographs, Seo was narrating not only an extremely reality-based story but also a little larger perspective. *Mottagi99* was an experiment on the rearrangement of densities and the form-making quality of architecture through the variations of formal language called "mottagi" (chamfer), but Seo herself declared it as a challenge to the urban architectural type of multi-family housing. Similarly, *Concrete Library*, *Project Re-interpret*, and *Archived Mass* not only were practical remodelling works within limited budgets, but also settled as the works of proposing sustainable strategies of the similarly formed buildings constructed at the same type of district zones in a similar period. How she proposed expandable general solutions through abstraction beyond the specific solutions reflected already in respective works was interestingly contrasted with the presentations of the other two winners, who discussed beauty through a series of episodes and showed how they had achieved the meticulous perfection of works through inner orders. Rather than careful devotion, Seo's presentation struck me as daring and bold strokes which I regarded as the young architect's vigor.

The clear, honest, and meticulous can be understood and sympathized with. However, what moves us greatly often comes from the incomprehensible.

Sooin Yang
Sooin Yang is an architect and public artist based in Seoul. His works range from architecture, and participatory art to marketing campaigns in content, and buildings to palm-sized devices in medium and scale. With the belief that the mixture of media can devise a more effective and inspiring solution to a particular problem, Yang aims to create 'things' with stories that can touch 'life' in a positive way.

Yang earned Bachelor of Architectural Engineering from Yonsei University, Korea, and Master of Architecture from Columbia University, where he was an Adjunct Assistant Professor and Founding Co-Director of the Living Architecture Lab. Since moving back to Seoul, Korea in 2011, he directs Lifethings/삶것, a multi-disciplinary design office.

심사총평 '젊기 때문'보다 '좋은 작업'

이민아 | 심사위원장, 건축사사무소 협동원 대표

젊은건축가상이 건축물이나 설계안이 아닌 건축가에게 수여하는 상이라는 점에서 매해 건축계는 어려운 공동 숙제를 정성스럽게 치르고 있다. 자신의 면허로 준공된 작품이 있는, 기준 연령에 해당하는 자에게 수여하는 상이지만 심사 과정에서 더 알아버린 건축가, 이른바 '사람'의 종합적 역량과 서사 때문에 심사위원들은 심사숙고에 빠졌다. 우리가 '건축가'로 정해놓은 심사 대상의 총체에 그들이 편집해 보여주는 이미지, 텍스트 너머 건축가의 정신도 포함된다면 그건 감히 기성 건축가가 어떻게 짐작하겠냐마는 심사위원들은 그 사이를 비집고 들어가 보려고도 애썼다. 46개 지원팀 중 절반은 이미 올해가 처음은 아니고 다회차 지원자들도 적지 않으니, 수상의 의미보다 도전 자체의 의미를 어떻게 인식하고 있는지 궁금했고, 혹시 지원 준비가 매해 자신의 주요 과제가 된 것은 아닌지 공연한 우려도 있었다.

46팀 포트폴리오들 간의 두드러진 대별은, 소규모 민간 프로젝트 위주 작업, 중소규모 공공 프로젝트 위주 작업으로 갈린다는 것. 그럼에도 프로젝트 수주에는 저간의 사정과 방식이 있고 건축가가 일관되게 천착하는 지점은 고유하게 드러나므로 심사의 그 어떤 기준도 발주 유형에 따른 프로젝트 여건과는 무관했다. 물론 공공 프로젝트 수행에는 대부분 시간이 설계를 위해서라기보다 심의, 인증, 보고 등에 소요되고 공사비, 시공자 수준 등 여건의 한계로 '이 정도로도 대단한'의 가점을 기대할 수 있겠지만, 그것은 부족함을 도리어 인정하는 격이다. 소규모 민간 발주 건축물과 같은 체급으로 비교하기 어려우니, 이는 공공건축상 제도의 역할이기도 하다. 건축가를 선정하는 이 상은 작품의 완성도 자체에는 다소 관대해도 좋은가 역시 당해 지원자 간의 상대평가에 의해 기준이 달라질 것이다. 또한 프로젝트 자체의 선량함과 공공성이 건축가 평가의 지표가 되지 않도록 유의했다.

올해 지원자의 공통된 특징은, 건축가의 관심이 '건축' 자체에 집중되어 있고 도시, 설치, 전시, 출판, 가구 등 건축 밖으로 시선을 확장한 건축가는 소수에 불과했다. 아직은 건축설계만이 가장 재밌는, 순도(?) 높은 건축가 집단의 등장이다. 또한 작업을 설명하는 서술구조가 매우 유사한데 거창하고 모호한 관념론이 아닌, 건축주를 어떻게 만나 어떤 각별한 과정을 거쳤는가를 소박한 문체로 길게 나열하는 설명 방식이 트렌드인 듯했고, 그러다 보니 새로운 결점이 드러났다. 필력은 별도의 능력이지만 텍스트가 비문일 경우 보여주는 것 전반에서 신뢰를 얻지 못했다. 결국 건축설계와 글쓰기는 주제, 구조, 공간, 디테일을 다 갖추고 있는 점에서 유사성이 있고, 특히 건축가를 선정하는 상이라 사유가 언어로 옮겨지는 단계에 개인적으로 큰 관심을 두었다.

여덟 팀을 선정하는 과정에, 심사위원 한 명의 표를 얻은 지원자들에 대해서도 무거운 토론을 이어갔고, 이 결정에 심사위원 개인 성향이 직접적인 영향을 미치지 않도록

General Review
'good-quality works' rather than 'the young'

Minah Lee | Jury President, Principal of Hyupdongone

Every year, the Korean architectural profession is carefully taking the difficult collective task of presenting the Korean Young Architect Award to architects, not to their completed buildings or design projects. It is given to such architects under an upper age limit as to have completed buildings with their own licenses, but this year's jury had to fall in a profound deliberation on the comprehensive competences and narratives of the so-called "persons," of whom we came to know better during the jury process. Provided the whole subjects to deliberate on that we designate as "architects" include their spirits beyond the images and texts edited by themselves, the jury endeavored to penetrate in between them, no matter how these established architects could dare guess what are there. Half of the 46 applicant teams are already not the first participants by this year, with a considerable proportion of them having applied many times. Thus, we were curious of how they were making sense of their applications rather than of award-winning itself, while for the caution's sake, concerned about whether their main tasks became caught in the annual application for this award.

The 46 submitted portfolios are distinctly divided in character, featuring small-scale private projects or small- or medium-scale public projects. Nevertheless, such project conditions by the types of commissioning was not concerned with any criterion for deliberation, because each project commissioned has its own style and circumstances and what architects consistently adhere to differ by their own characters. Given a public project delivery consumes most time on deliberation, certification, and report drafting rather than on design itself and even is limited in its construction cost and the average level of trade skills, one might expect some bonus points taking such limitations as an excuse like "great enough for this much" but this is just admitting such drawbacks. Hardly compared to the small-scale private building projects at the same level, it must be addressed within the role of a public architectural awarding system. Whether this award presented to architects can allow some leniency in the perfection of their works will also be determined by different criteria depending on the relative evaluations among the applicants of the year. Also, we were mindful not to use the good intentions or public contributions of projects themselves as the indicators of evaluating their architects.

The applicants of this year are commonly characterized by the focus of architect's concerns on "architecture" itself. With few of them extending their perspective outside architecture such as the city, installation, exhibition, publication, and furniture, they evidence an emergent group of highly "pure" architects who are still most interested only in architectural design. Also, they explained their own works in a very similar narrative structure, which is not a grand or ambiguous discussion of rarefied ideas but

상의 의미를 지속적으로 짚었다. 새로운 세대의 특성이 전면화된 건축이어서가 아닌, 건축적 사유와 통찰, 실천력이 뛰어난 작가 중 기성의 규범에 편입하지 않은 채 확고히 자신의 언어를 다루고 있는 신인을 선정하는 상이라는 것이 맞는 표현이겠다. 즉 '젊기 때문'보다 '좋은 작업'이 압도적 전제였다. '발굴'과 '육성'이 상의 목적이던 시기도 있었지만, 14회 여덟 팀의 포트폴리오에서 우리는 2023년 가장 뜨거운 설계도면을 보았다. 이미지에 현혹될 걱정이 없었던 것은 건강하게 그려진 도면 덕이다. 상의 흥행, 수상자의 다양성 등이 심사 중 잠깐씩 언급되었지만 정작 수상자 세 팀은 비슷한 점이 있다. 도면은 왜 그리는가, 건축물에게 시간은 무엇인가, 나는 왜 이 얘기를 하려는가 등의 질문들로 자신과의 싸움이 지독했을 것이다.

서자민(아지트스튜디오 건축사사무소)
문제적 도시 현상을 설계 의도를 생성하는 결정적 단서로 단호하게 포섭한다. 그리고 곧장, 너무 '기본적'이라 질문을 게을리해왔던 문제들, 덩어리, 구축, 비움, 양감, 질감, 형태들의 개념을 향해 즐겁게 공격하며 설계로 이행한다. 심사위원이 보기에 즐겁고 과감한 결정일 테고, 정작 서자민 본인에게는 유희도 시도도 아닌, 기본의 기본에 천착하는 내면의 본능적 씨름이었을 것이다. 자신의 질문에 명료한 입장을 제시하는 작업 과정은 젊은 건축가의 고유한 내러티브를 넘어, 기성 건축계의 어쩌면 빈궁한 담론에 반한 원초적 물음을 던진다. 특유의 조형성을 못생김의 미학으로 언어화하여 집착을 덜어내는 태도는 건축가 내면에서 충돌하는 논리와 형체의 간극을 보여준다. 건축가의 직관적인 미감이 건축물 형태에 마지막으로 개입하는 것은 당연하고 어쩌면 소위 완성도라 일컫는 지점을 끝내는 무기가 될 텐데, 미학적 측면엔 초연했다는 발언은 아쉽다. 어떤 면에서 가장 감각적인 화면의 포트폴리오를 제출한 지원자로 젊은 건축가에게 시대가 기대하는 덕목에 이 또한 결코 벗어나지 않기 때문이다.

김진휴+남호진(건축사사무소 김남)
건축에서의 아름다움을 심사에서 질문했고 답을 들었다. 이 일이 벌어진 것만으로 두 사람 건축은 의미있다. 추천 아닌 자천으로 참가하는 상의 성격상 고도의 자기예찬 전략이 필수라면 두 사람은 별로 신경을 못 쓴 듯 도리어 모두가 회피하는 건축의 아름다움을 읊었다. 아름다움이 좁은 문을 통과하는 적확한 구조와 치밀한 기술적 해결, 엄격한 시각적 완성도에 의지하는 개념임을 환기시키면서도 표피적이고 관성적인 결정은 끼어들 틈 없음을 강조한다. 즉 눈에 보이지 않는 것으로부터 정의하고 치열하게 구조화하여, 시공자의 수고, 사용자의 기쁨, 건축가 스스로의 검열이 동반될 때 비로소 아름다움에 이르는 길에 당도했음을 성찰하는 태도로 '결국 건축가는 무엇에 헌신하는 사람인가'라는 질문을 강력히 던지고 있다. 두 사람의 건축은 인간과 주변을 사색하는 일에서 시작하여, '아름다움'보다 실은 훨씬 어려운 '윤리적' 건축의 실천을 위해 부단히 노력 중임을 목격했다.

a long description of how they met their clients and what special process they have undergone, a structure that seems to be a trend among them. However, this revealed a new drawback. While writing is a distinct skill separate from architecture, any text including ungrammatical sentences failed to instill trust in the overall presentation of the architect. After all, architectural design and writing share similarities, as both involve themes, structures, spaces, and details. Personally, I was particularly intrigued by how applicants translated their thoughts into language because this award is presented to architects themselves.

In the process of selecting 8 finalists, we had serious discussions even on the applicants who received only one juror's vote, and to ensure that the jurors' personal preferences did not directly influence this decision, we consistently reminded ourselves of what this award signifies. It was not about all-out features of a new generation, but would be rightly put to select the fresh blood dealing with their own language resolutely without being accustomed to the established conventions, amongst those with great architectural thinking, insights, and practical abilities. That is to say, the significance of this award is dominantly premised on selecting "good-quality works" rather than the "young." Although this award was once oriented to "hunting" and "fostering" new talent, this year's 14th award showed the most robust drawings from the 8 finalists' portfolios. Thanks to their sound drawings, we had no worries about being misled by images. Even as how to lead this award into popular success or the diversity of winners was briefly remarked during the jury process, the three winners shared similarities indeed. Their struggling with themselves would have been tenacious, grappling with such questions as "Why do we make drawings?", "What does time mean for a building?", and "Why do I intend to talk this?"

Jamin Seo (Agit Studio)
She encapsulates a problematic urban phenomenon sternly with a decisive clue that creates a design intention. Straight away, she enjoyably strikes home of the issues that have been too "basic" concepts to be profoundly questioned, such as massing, construction, emptying, sense of volume, texture, and forms, while turning them into design. While this was probably seen by the jury as a joyful and bold decision, it would have been neither a play nor an attempt for Seo herself but just her instinctive inner struggle delving into the very basics. The process of presenting a clear position on her own questions goes beyond the young architect's unique narrative, posing fundamental questions that challenge the impoverished discourse of the established architectural community. The attitude of articulating a unique formalism into the verbal expression "aesthetics of ugliness" reveals the gap between the logic and form conflicting inside the architect's mind. It is natural for the architect's intuitive sense of beauty to finally intervene in the form of the building, perhaps becoming the weapon that ultimately completes the so-called degree of perfection. However, her statement of aloofness about the aesthetic aspect is somewhat regrettable, because her portfolio was in a way the most sensuously visualized among all applicants which can never be outside the virtues of a young architect expected by the era."

김영수(모어레스 건축사사무소)

재료 본연의 성질, 건축 요소의 자리, 사물과 공간의 관계 등 건축 본질에 대한 집요한 탐색과 사유 과정이 건축가 개인의 지적감수성(intellectual sensitivity)에서 비롯된 태도를 넘어 젊은 건축가상의 당대성과 미래적 역할에 방향타가 될 만한 유의미한 모습으로 전달됐다. 심사를 하면서 과연 이들에게 건축이 고통스럽지 않고 즐겁다면 지속가능한 내적 영역이 강건하게 있는가를 추측해 보았고, 대표적으로 김영수가 그랬다. 엄격하게 조정한 공간의 형태, 재료의 두께, 빛의 강도를 현장에서 섬세하게 통솔하여 완성도 높은 건축물로 실현해가는 과정 속에 건축가 특유의 오기와 환희 그리고 스스로를 향한 비평이 즐겁고 균형 있게 자리 잡혀 있음이 엿보였다. 특별히 '단면적 공간'을 지극히 신중하게 해설하는 젊은 건축가를 만난 것은 본질론에서 이미 멀리 떠난 채, 건축의 힘듦에 대한 궁색한 핑계거리들로 보호받고 싶어했던 기성건축가로서 행운이다. 심사위원들은 전원 일치로 김영수를 올해의 주목할 건축가로 선정했다.

2023 젊은 건축가상 심사위원

심사위원장	이민아(건축사사무소 협동원)
심사위원	김동진(로디자인, 홍익대학교)
	양수인(삶것건축사사무소)
	이규상((주)보이드아키텍트 건축사사무소)
	정웅식((주)온건축사사무소)

Jinhyu Kim + Hojin Nam (KimNam Architects)

We questioned and received answers about beauty in architecture from this duo. This discussion alone is enough to make their architecture meaningful. Considering the nature of this award, receiving self-recommendations of applicants who thus essentially need a high level of self-promotion strategy, the duo seemed to care less and instead recited the beauty of architecture that everyone tends to avoid. Reminding us that beauty is a concept that relies on precise structures, meticulous technical solutions, and strict visual perfection, they emphasized that superficial and inert decisions had no room to intrude. In other words, they reflected on how they were able to reach the path to beauty only when making definitions and rigorous structuring with the invisible, accompanied by the constructor's efforts, the user's joy, and the architect's self-censorship. In this attitude, they are strongly posing the question, "Ultimately, what is it that an architect dedicates themselves to?" We witnessed the duo's architecture begins with contemplating human beings and the surroundings and is tirelessly striving for practicing "ethical" architecture, which is actually far more challenging than realizing "beauty."

Youngsoo Kim (More Less Architects)

His persistent exploration and reasoning process on the essence of architecture, such as the inherent properties of materials, the placement of architectural elements, and the relationship between objects and space, was so significantly conveyed as to become a rudder that could guide the contemporaneity and future role of the KYAA beyond the architect's individual intellectual sensitivity. During the jury process, I speculated whether these architects if architecture is not distressing but enjoyable for them have an inside strong enough to be sustainable, and found its exemplar from Youngsoo Kim. In the process of strictly adjusting the form of spaces, the thickness of materials, and the intensity of light and meticulously orchestrating them on site into a building in great perfection, he revealed a delightful and balanced presence of the architect's peculiar passion, joy, and self-critique. Encountering a young architect who interprets "sectional space" with extreme care is fortunate for established architects who, already having moved far away from the essence, sought refuge in lame excuses for the difficulties of architecture. The jury members unanimously selected Youngsoo Kim as the noteworthy architect of this year.

Jury of the Korean Young Architect Award 2023

Minah Lee(Hyupdongone)	President
Dongjin Lee(L'EAU Design, Hongik University)	Juror
Sooin Yang(Lifethings)	Juror
Kyusang Lee(VOID Architects)	Juror
Woongsik Jung(On Architects)	Juror

의미, 무용, 태도		Meaning, Futility, Attitude
2023 젊은 건축가상		Korean Young Architect Award 2023

김진휴, 남호진
김영수
서자민

Jinhyu Kim, Hojin Nam
Youngsoo Kim
Jamin Seo

발행일:	2024년 2월 26일 2쇄 발행
저자:	김진휴, 남호진, 김영수, 서자민
발행인:	정귀원

Published on 20, December, 2023
Authors: Jinhyu Kim, Hojin Nam, Youngsoo Kim, Jamin Seo
Publisher: Kwiweon Chung

발행처:	제대로랩
	등록번호 제2018-000047호
	서울시 종로구 사직로8길 15-2 4층
	전화 02-2061-4146
	zederolab@gmail.com

Published by zederolab
4F, 15-2, Sajik-ro 8-gil, Jongno-gu, Seoul
T.+82-2-2061-4146
zederolab@gmail.com

기획·편집:	제대로랩
편집디자인:	강주현
진행:	임선희(새건축사협의회)
번역·감수:	조순익
인쇄·제책:	세걸음

Planning & Editing: Zederolab
Design: Juhyun Kang
Assistant: Sunhee Lim
Translation & Proofreading: Soonik Cho
Printing & Binding: Seguleum

ISBN: 979-11-965929-7-4(93540)

ISBN: 979-11-965929-7-4(93540)

이 책에 실린 모든 글과 이미지는 저작권법에 따라 보호받으며 어떠한 형태로든 무단 전재와 복제를 금합니다. 정가는 뒤표지에 있습니다. 잘못된 책은 구입처에서 교환해 드립니다.

All rights reserved. No Part of this book may be reproduced in any form or by any means, electronic or mechanical, without the prior written permission of the copyright holder.

본 사업은 국민체육진흥기금의 후원을 받아 시행하는 사업입니다.

문화체육관광부

후원:	볼라코리아,
	(주)삼익산업,
	에너지엑스(주),
	디자인후즈,
	(주)이안알앤씨